# *Magic Enzymes: Key to Youth and Health*

*Carlson Wade*

Parker Publishing Company, Inc. West Nyack, N.Y.

*To Your Renewed Youthful Health*



**Library of Congress Cataloging in Publication Data**

Wade, Carlson.
Magic enzymes.

Includes bibliographical references.
1. Nutrition. 2. Enzymes. 3. Rejuvenation.
I. Title. [DNLM: 1. Enzymes--Popular works.
2. Nutrition--Popular works. QU135 W119m 1973]
RA784.W258 615'.854 73-9830
ISBN 0-13-543942-6

Printed in the United States of America

# *Foreword*

## *by a Doctor of Medicine*

The book you hold is the key that unlocks Nature's secret doorway to the Fountain of Youth. This book takes you into Nature's own healing garden of health. It shows you how to use Nature's own plant medicines that help you restore and rejuvenate your mind and your body from head to toe.

Carlson Wade, the renowned writer on medical subjects, tells of Nature-created substances—the *enzymes*—found in foods as close as your own pantry or refrigerator. He describes how these foods can be easily used to help give you the glow of youth and the feeling of radiant health. These plant substances provide rejuvenation from the foods you eat. The magic enzymes can be used to help give you a "forever young" skin, promote youthful flexibility in your arms and legs, and give you more "go power" vitality and energy.

Mr. Wade has collected hundreds of all-natural magic enzyme secrets from Europe, the Pacific, and the tropics, as well as from nations of the past and present throughout the world. He reports on physician-prepared programs using magic enzymes for "miracle" rejuvenation healing. These programs are outlined in simple step-by-step form. You can follow them right in your own home. Many take only moments to prepare with ingredients found in your own kitchen or available at almost any corner food market at a modest cost. These easy but effective all-natural programs offer hope for your rejuvenated health, thanks to their magic enzyme power.

Throughout the chapters of this book, Mr. Wade offers you programs for almost any kind of health situation. He reports, for example, on the use of fruit enzymes to rejuvenate your digestive system, on how vegetable enzymes help give your skin a soft-as-silk feeling, and on how to follow the Swiss raw food plan with "Youth Enzyme X" for self-rejuvenation. He describes the dynamic youth power of seed sprouts and tells you how to grow these magic enzyme foods in a glass jar on your kitchen shelf. He tells you how to use ordinary water programs to alert and awaken sluggish body enzymes and feel a supercharging of youthful vitality from head to toe.

Carlson Wade offers more than the listing of a treasury of secrets on using Nature's plant medicines and enzymes for the health and vitality of your body and mind. He explains clearly just *how* you can use these simple home programs for renewing your youthful health. Then he explains *why* these home programs hold the key to a "forever young" vitality for your body and mind, and how to let these magic enzymes work for you.

This is a thoroughly researched and documented, highly practical book that is the key to rejuvenated youthfulness. The book is informative, complete and easy to use. Its information is presented in such a simple step-by-step manner that just about everyone can learn how to use the programs to enjoy natural healing and vigorously youthful health. The documented case histories presented in connection with the programs offer hope for prompt health rejuvenation, thanks to Nature's miracle workers, the *enzymes.*

William S. Keezer, M.D.

# *What This Book Can Do for You*

As a medical reporter and professional editor, I discovered many scientific papers describing certain substances in everyday foods and plants that could be used to create "miracle" healings and restoration of youthfulness.

In particular, researchers told of a "magic" element found in raw plant foods that could help boost a feeling of sparkling youthfulness and slow down the aging process. From all parts of the world, medical doctors, pharmacologists, and laboratory scientists reported discoveries of the rejuvenating and curative potential of all-natural plant food substances known as *enzymes.* The successful rejuvenation properties contained in these food substances were described in many medical, scientific and nutritional journals. Many hailed the enzyme as the "missing link" for a youthful long life. Indeed, the modern scientific reports of healing and youth regeneration make enzyme powers an exciting discovery that may well be Nature's secret of more youthful health at any age.

When I began researching the secrets of using food enzymes as natural healers and as means of rejuvenation, it became apparent that there were a great many people who needed to be informed about a completely natural health program that would give them the healthful vitality they could not obtain from chemical or drug therapy. These persons wanted to rejuvenate their digestive organs, enjoy supple limbs and flexible muscles, radiate the glow of health, and enjoy the fresh bloom of youthfulness at any age. They needed to know the miracle rejuvenation powers of enzymes in simple and easy programs prescribed by doctors and other health authorities the world over. To share this knowledge meant that I would have to devote several years to intensive research and writing. In so doing, I consulted with the leading medical doctors, pharmacologists, herbal pharmacists, and scientists from America,

Europe, Asia and various island nations from the Atlantic, the Pacific, and the Caribbean. Many translations had to be made. The scientists consulted very generously shared their knowledge of enzymes with me so that I could put everything together in one volume to offer a "head to toe" program that would help revive-restore-rejuvenate your health. This book is the product of over a decade of such intensive enzyme research, complete with all known reported case programs and step-by-step methods for using enzymes for your rejuvenated health.

This book will benefit all of you who want to use *natural, drugless,* and *medicineless* programs to help promote your healing rejuvenation as well as better measure of health. All of the enzyme-bearing foods in this book are available at almost any local food market as well as at health food shops.

Most of the enzyme programs in this book entail no extra cost to you. This book can well be your key for more youthful health at any age through *natural* methods using *natural* enzyme foods.

Carlson Wade

**Other Books by the Author**

*Helping Your Health with Enzymes*
*Magic Minerals: Key to Better Health*
*The Natural Way to Health Through Controlled Fasting*
*Carlson Wade's Gourmet Health Foods Cookbook*
*Natural and Folk Remedies*
*The Natural Laws of Healthful Living*
*Health Tonics, Elixirs and Potions for the Look and Feel of Youth*
*Natural Hormones: The Secret of Youthful Health*
*Health Secrets from the Orient*

# Contents

# 1

# How Natural Food Enzymes Hold the Magic Key to a "Forever Young and Healthy" Body and Mind

The secret of looking and feeling "forever young and healthy" lies within your own body. Nature has placed within your system a set of special substances that have a "magic" power of perpetual healing. These magic-like Nature-created substances are found in every body gland, cell, tissue, organ, and process of metabolism. These substances influence every physical and mental function from fluttering your eyelids when you awaken in the morning to stretching out to sleep at night. To enjoy a feeling of youth and health of your body and mind, you need the "life force" that is sent forth by these "magic" substances in your body. They will help give you miracle protection against the onslaught of aging. They will help give you those healing properties that enable you to recover from the stresses and strains of everyday living. These Nature-created "magic" substances are known as *enzymes.*

## *How Enzymes Help Promote the Look and Feeling of Youthful Health*

**What Enzymes Are.** Enzymes are complex protein-like biological substances found in all living cells, whether animal or vegetable. They are magic-like substances that enable you to digest food and absorb the life-giving and youth-perpetuating nutrients through the process of assimilation. These protein-like life principles are intimately involved in the action and activity of every cell in the human body and in every function of life.

Enzymes are substances that turn a slice of apple brown when it is exposed to the air. They give cheese its flavor. They convert food to youthful energy, build and rebuild cells and tissues to help repair and regenerate the body from the inside to the outside, help send a stream of nourishing ingredients to the bloodstream which uses them for "feeding" almost all glands, organs, cells, and tissues as well as the brain itself. Enzymes are "youth substances" that extract nutrients from foods in the digestive tract and use them for building and rebuilding the body from head to toe. Enzymes are intimately involved in almost every single body and mind function. Life could not exist without enzymes. Youthful appearance of the body and youthful alertness of the mind could not be possible without enzymes–the magic keys that unlock the secret of perpetual youth and health.

**How Enzymes Work in a Catalyst-Like Youth-Building Process.** Enzymes in your body act as *catalysts.* That is, they create biological changes in foods to prepare them for better digestion and assimilat n. But enzymes themselves do not enter into the processes. As an example–fire cooks food that you eat. But you do not eat the fire itself. You eat the food that has been cooked by the fire. So it is with the catalyst-acting enzymes. They prepare food in your system to be used for the building and rebuilding of your entire body and mind. Without these enzymes, or with a weak enzyme system, food that you eat would remain undigested as a lump, or they would be only *partially* digested. This means your body may be receiving a good amount of nourishing food, but it is not assimilating or digesting this food and is being denied full youth-building nourishment.

*Secret of Enzyme Youth Power:* What you eat is not as valuable as what you assimilate! To help perpetuate the look and feel of youth, enzymes must be strong, vigorous and plentiful in order to take out all of the youth-building nutrients in food that you eat. A strong and healthy enzyme system holds the key to "youth power" at all ages.

**How Enzymes Help Create Youthful Health.** Here are some of the youth-building benefits of having a strong enzyme system. Enzymes help perform the following regenerative processes right within your body–Enzymes rebuild healthy food into youthful muscle, nerve, bone and gland. Enzymes help store food nutrients in your liver and muscles for future youth-building purposes.

Enzymes are involved in the formation of urea and other wastes to be eliminated in perspiration and urine. Enzymes are used to eliminate unnecessary carbon dioxide from the respiratory system and help "air wash" the lungs.

Enzymes use phosphorus and other minerals to build and rebuild strong nerves and bones. Enzymes help fix iron in the red blood cells. Enzymes help bring about coagulation of the blood to promote better healing and less bleeding. Enzymes help decompose age-causing hydrogen peroxide and liberate needed oxygen from it. Enzymes promote the youth-building process of oxidation; that is, a union of oxygen with other substances to give the "breath of life" to each and every body cell and tissue.

Enzymes attack waste materials in the bloodstream and tissues and transform them into substances that can be eliminated from the system. Enzymes change food substances into nutrients that can then be taken up by the body processes and used for the life processes that help build youthful health from head to toe. As catalysts, enzymes are the body's spark plugs. They create a healthful chain reaction that gives the body its zest and vigor. They hold the key to youth and health.

**Different Types of Enzymes.** There are over 700 different types of enzymes. Each enzyme performs a specific function. That is, each enzyme has one individual task. One enzyme converts food into energy. Another enzyme transforms proteins. Still another enzyme works on carbohydrates and still another enzyme works on the different vitamins and minerals and fats. There is an enzyme for each and every specific body function.

**How Enzymes Enter Into Youth-Health Processes.** Your body is composed of billions of minute units known as cells. Each cell is a living individual unit connected with other cells to make up your body. To help perpetuate the youth-health process, each individual cell must digest food, excrete its wastes, repair itself and be ready in a fraction of a second to perform its specialized task. These cells depend upon enzymes to enable them to perform their assigned youth-health processes.

*Every cellular reaction is dependent upon a biological response triggered and controlled by a specific enzyme system. All youth and health* processes consist of a great number of these enzymes working together in coordination. *Youth and health are influenced by the protein of the cell joining with a particular vitamin and nutrient to form an enzyme that promotes a valuable function.*

**Magic Power of Enzymes.** Every cell in the body must contain up to 100,000 of these enzyme systems ready to go to work in a fraction of a second. They help defend you against infection, they help in the process of assimilation of food, they help in rebuilding your bloodstream and in regenerating your body from the inside to the outside.

**Enzymes Require Nourishment.** Valuable vitamins, minerals and proteins must be in each body cell at the same time and in proper amounts if the enzyme system is to be properly nourished and able to create its youth-building powers. A well-nourished enzyme system is able to help repair the wear and tear of body cells and tissues. The body that is made out of these cells and tissues can now enjoy youthful health, thanks to a well-nourished enzyme system.

### *12 Magic Enzymes and How They Help Perpetuate Youthful Health*

While each and every enzyme is vital, there are 12 that stand above all in being able to help perpetuate a feeling of youthful health. Here are those 12 magic enzymes and their youth-building functions:

1. Ptyalin

*Location:* Mouth–salivary glands.

*Function:* Acts upon carbohydrates and starches.

*End Result:* Creates a form of dextrin and maltose to create energy-producing benefits.

2. Pepsin

*Location:* Digestive system.

*Function:* Acts upon protein foods.

*End Result:* Creates forms of amino acids such as peptones and proteoses to be used by the bloodstream, bones, glands, cells and tissues for nourishment and rebuilding of the body.

3. Rennin

*Location:* Digestive system.

*Function:* Acts upon dairy products.

*End Result:* Helps to coagulate and change casein (milk

protein) into amino acids for body rejuvenation. Rennin also metabolizes dairy minerals (calcium, phosphorus, magnesium and others) and sends them via the bloodstream to help build your bones, teeth and arterio-neural systems.

4. Lipase
*Location:* Digestive system.
*Function:* Acts upon fats in dairy products.
*End Result:* Helps to split the fat portion of dairy products into forms that can be used by the cells and organs for basic health and for transportation-storage of vitamins and minerals.

5. Trypsin
*Location:* Intestinal-pancreas region.
*Function:* Continues the digestive work of dairy proteins after the preliminary work has been done by pepsin and rennin.
*End Result:* Helps to take out amino acids as well as vitamins and minerals for the use of body cells, tissues and organs throughout your system from head to toe.

6. Steapsin
*Location:* Intestines.
*Function:* Acts upon fats from dairy foods.
*End Result:* Metabolizes these fats and transforms them into usable essential fatty acids that are needed to lubricate the arteries and maintain a healthy and youthful bloodstream.

7. Amylopsin
*Location:* Intestines.
*Function:* Acts upon starches in foods.
*End Result:* Works to assimilate starches from your bread products to create a glucose-energy byproduct that gives you "go power."

8. Invertase
*Location:* Intestines.
*Function:* Acts upon sugars in most foods.
*End Result:* Helps to metabolize natural sugars in foods into forms of glucose and fructose which are then stored in

body organs to be made available as natural energizers for your body and thinking processes.

9. Maltase

*Location:* Intestines.

*Function:* Acts upon bread and grain carbohydrates.

*End Result:* Helps to transform the ptyalin-digested carbohydrates into maltose which then creates dextrin, a form of energy-creating natural sugar.

10. Lactase

*Location:* Intestines.

*Function:* Works upon the milk sugar in dairy products.

*End Result:* Helps to transform milk sugar (lactose) into forms of glucose and galactose which can then be used to create better energy and vitality and a more vital assimilation process throughout the body.

11. Hydrochloric Acid

*Location:* Intestinal-digestive Systems.

*Function:* Works upon heavy protein to extract amino acids for body building blocks. Propels vitamins and minerals to perform health- and youth-building functions throughout the body.

*End Result:* Takes up amino acids as well as vitamins and minerals to be transported via the bloodstream to main body glands and cells and tissues.

12. Erepsin

*Location:* Intestines.

*Function:* Acts upon protein and peptides in foods.

*End Result:* Metabolizes protein into usable amino acids for the healthful regeneration of the entire body, internally and externally.

**Enzymes: The Key to Extended Youth.** So we see that enzymes work to convert food into youthful vitality-energy, helps build health into body cells and tissues. Enzymes take all the nutrients such as vitamins, minerals, proteins, carbohydrates, and fats and transform them so they can be used to help build strong muscles, strong bones, smooth skin, healthy hair, youthful glands and healthy hormones. Furthermore, enzymes help protect you against inflammations, speed up the healing processes, and help to protect you against pollution and chemicals in our environment. They help promote better removal of waste products such as urea and carbon dioxide through your body pores and natural channels. They also help in blood coagulation, balance the youthful oxygenation of blood and cells. They are involved in helping give you youthful stamina and a stronger libido, help keep you thinking young and boost your powers of mental vigor. Enzymes are tiny but they are powerful in their encompassing youth-building functions.

**Each Enzyme is Individual.** So we see that an enzyme (or ferment) is a protein substance produced by living cells which promote youthful metabolism. Each enzyme is individually responsible for only one type of change so it is essential to be adequately nourished to have full use of all enzymes. A deficiency of one enzyme can cause malfunctioning of the entire chain reaction. Enzymes are highly specific: one enzyme will catalyse only one kind of reaction and no other! Furthermore, many enzymes are weak until they join other enzymes. *And all enzymes depend upon adequate nourishment to function. Enzymes must be fed!* See that they are nourished and they help perform their individual and powerful youth regenerating processes.

## *How Nutrients Help Boost Enzyme Vigor*

Formation and activity of body enzymes require a balance of nutritional substances. Your mouth enzymes require minerals to remain alkaline. Your digestive system requires Vitamin $B_1$ (thiamine hydrochloride) to promote enzymatic digestion of protein in an acid environment. Your intestinal enzymes require Vitamin $B_2$ (riboflavin) to join with protein to promote utilization and function. Most of the enzymes require goodly amounts of Vitamin C since they hasten the combining of oxygen with this vitamin in natural processes. The other vitamins and minerals are needed to promote a balance. The activity of some enzymes will use up lots of Vitamin A so this nutrient must also be present in sufficient quantities to obtain maximum youth-building benefits from these wonder workers of the body. It is essential to eat a well-balanced diet of wholesome, natural foods, to give your system the working materials out of which enzymes fashion those ingredients that enter into the building of a youthful body and mind.

### *Where are Enzymes Found?*

Your body creates many enzymes from substances (food, liquid, air) that go into your body. These substances help make many enzymes. There are two basic sources of enzymes:

1. *Endogenous Enzymes.* Your body creates these enzymes from raw foods, liquids and air that you breathe. These are enzymes that are found within your body.
2. *Exogenous Enzymes.* These are enzymes that are found in raw foods. Such enzymes are powerful protein-like substances that are brimming with vigor to help give your body the vital youth-building elements required for good health.

*Note:* All Nature-created foods (from the ground, orchard, or field, or from animal sources) contain enzymes. But any form of cooking will destroy them. Hence the natural law of enzymes for youth and health: *eat raw foods for magic enzymes!*

### *How Cooking Destroys Valuable Enzymes*

Enzymes are found *only* in raw foods. The foods must be natural, nonprocessed, organic. Cooking, heating, bottling, canning, pasteurizing or processing will *destroy* valuable enzymes. Temperatures over 122° F will kill enzymes. If you cook your fruits, vegetables or other foods, it means that the heat kills off the valuable enzymes. So you would need to adjust your eating programs and emphasize raw foods. It is true that many foods such as meat, poultry, fish, eggs and many beans and grains, are unpalatable if eaten raw. But it is also true that there are many foods that are much tastier, crunchier, healthier and more delicious when eaten *raw.* In particular, fruits, vegetables, many nuts, seeds, and some grains, can all be eaten raw for their prime enzyme power. You reap more benefits from raw foods which are prime sources of vitamins, minerals, essential fatty acids and types of proteins that are easier to digest if uncooked. Cooking will destroy not only enzymes but many of these valuable nutrients. So you see that fresh, nonprocessed raw foods are your prime sources of good nutrition and youth-building enzymes.

### *How Five "Everyday Foods" Can Give You Abundant Enzymes for Youthful Health*

The name *enzyme* comes from the Greek *enzymos* which means *leavened* or *fermented.* Long ago, man recognized that fermented foods promoted a feeling of health and youth. Fermented foods are *raw foods* with a very special "secret of enzyme youth." The process of fermentation helps create a natural "acid" environment in the food that is especially hospitable to enzymes—both in the food and in your body. The simple process of fermentation causes a natural multiplication and increase of food enzymes. This process then magnifies and boosts the power and health of the enzymes, giving them a dynamic source of youth building elements. In the past, man often fermented foods to keep them fresh. He thus enjoyed a long life that was almost always free from illness. This primitive man knew that fermented foods held the key to everlasting youth. He ate raw foods as well as fermented foods and enjoyed extended youth and virility well beyond the century span. Today modern science recognizes that fermentation will help boost the vigor of food enzymes. Eating or drinking fermented foods with such enzymes further helps stimulate the body's own often sluggish chain of enzymes. Once alerted, thanks to fermented foods, enzymes can go to work in their metabolism to promote youthful and healthful living.

Here are five "everyday foods" that can be fermented at home, or purchased ready-made at most health food stores. They are prime sources of enzymes. They are prime sources of "activators" that help alert your own sluggish enzymes. In this harmony, they are wonder foods for youthful living.

1. *Buttermilk.* Anyone who has overbeaten whipping cream has an idea of how butter is extracted from cream. As the chilled cream is beaten, the fat globules begin to clump

together. Soon the smooth cream separates into light yellow creamy butter and a thin liquid called buttermilk. This is a process of fermenting or "souring" in which the butter particles shed their thin coating into the buttermilk. The coating of complex phospholipid material is the end result of the fermenting–and a powerhouse of rich enzymes. A further benefit is that the valuable enzymes in buttermilk help introduce proteins and vitamins as well as minerals into the digestive tract, thus favoring a healthy environment for better assimilation.

*Enzyme Booster:* Each morning, drink one glass of fresh buttermilk to help "churn up" sluggish digestive enzymes and prepare your system for the day's intake of food. One glass does much to help promote better digestion for your approaching breakfast. Enzymes will take nutrients from your ingested breakfast and use them for promoting energy and mental vitality through regulated blood sugar levels.

2. *Cabbage.* This raw vegetable has an amazing supply of vitamins, minerals and enzymes. To ferment, soak raw cabbage in a glass jar of water and sea salt. Then eat the raw cabbage leaves and mix the liquid with other fresh raw vegetable juices for a rejuvenating pickup that helps you look and feel good for much of the day. A secret benefit here is that fermented cabbage is a prime source of auxin ingredients which boost the activity of enzymes and help nourish them to create better vigor and health.

*Enzyme Booster:* Before beginning your main meal, eat a raw fermented cabbage salad with its juice as your dressing. This helps saturate your system with needed enzymes that are waiting to promote better digestion for the meal ahead.

3. *Sauerkraut Juice.* Here is a tart, acid beverage that creates a natural flow of hydrochloric acid, needed to promote better digestion and youthful metabolism. Sauerkraut juice was well known to many Europeans who enjoyed long life and youthful vigor by drinking a glass or two daily. Modern processed sauerkraut juice, however, is inadvisable to use since it is chemically preserved and processed and is contrary to body health or enzyme health. Furthermore, commercial sauerkraut juice is devoid of enzymes since

heating and processing destroy these magic youth builders. Here is how you can make your own (fermented) sauerkraut juice:

Use a large earthenware jug, a glass jar or a wooden barrel. (Do NOT use aluminum or any metalware.) Pack sliced white cabbage into the jar. Tamp down with a wooden masher until the juice rises above the cabbage. You may add water to get this result. Sprinkle a few desired herbs on top. Add several strips of raw vegetables. Add just a sprinkle of sea salt. Cover with more grated cabbage leaves until the jar or container is filled. (Be sure to press down hard so the cabbage can be saturated with its own liquid. This aids in fermentation.)

When filled, cover with a clean cloth topping over which you put a wooden board. Put in a warm place where the temperature is between 60°F to 70°F but no higher or lower. This temperature control is necessary to enable the enzymes to create fermentation and produce the best type of nutrition-rich juice. Let stand about three weeks. You may need to remove the cloth topping for washing at intervals to eliminate the mold that would otherwise penetrate the cabbage.

When ready, store in a cool place or put in glass jars in your refrigerator.

*Enzyme Booster:* Drink sauerkraut juice "straight" or mix with tomato juice. This fermented juice is reported to be a powerhouse of enzymes that works to rejuvenate,awaken and revive your body enzymes. Drink a glass between meals or whenever you feel the need for a pickup or energy booster. It is a European "secret" for youthful health because its fermented juices are brimming with healthy, youthful and powerful enzymes.

4. *Whey.* The nursery rhyme about Little Miss Muffet who sat on a tuffet (low seat) eating her curds and whey may well hold the clue to better health through enzymes. Whey is a liquid portion of cottage cheese. It is believed that whey is to milk what plasma is to the bloodstream. Whey is the serum which remains after the coagulation of casein when cheese is made. Whey is a prime source of enzymes along with lactose, lactalbumin and many valuable minerals. Whey enzymes are

exceptionally nutritious because whey is a soured milk byproduct. Whey is a fermented milk that sends enzymes into the system to promote a youth-building function. Namely, enzymes send lactobacillus organisms into the digestive-intestinal system to create an environment that is self-cleansing. This environment works to cleanse the colon and rid it of infectious bacteria.

*Unique Benefit:* Fermented whey sends enzymes into the system to break down lactose (a milk sugar) in the intestine into lactic acid. Whey enzymes then use these lactic acids to cleanse the colon, to send minerals into the absorption process and hasten the absorption of lactose. This lactose enzyme nourishes the small and large intestines, creates an environment for healthful acid-forming ingredients. Lactose enzymes produce a degree of healthful colon acidity which is hostile to putrefactive or disease-causing bacteria. Lactose enzymes thus help establish lactobacilli which make it difficult for harmful staphylococci to take charge. This is the natural way to maintain healthful intestinal homeostasis or balance. Lactose enzymes also bring about better assimilation of calcium, phosphorus and magnesium along with other minerals and vitamins. Whey (fermented milk) is 70 percent lactose which immediately goes to the colon to create this healthful environment. Little Miss Muffet was diminutive in body but wise in thought to know the value of this fermented milk.

*How to Make Whey:* Use an enamel or earthenware container. (Never use aluminum or metalware.) Add a dollop of yogurt to a container of milk, preferably safe (certified) raw milk. You can use pasteurized milk too. Let it stand covered with a cheesecloth up to 24 hours until the milk is thick and the whey liquid separates out. Stir with a spoon. Gradually separate the curds (lumps) from the whey. Drain the whey into a clean jug. (You may eat the curds as a cheese.) This gives you a good supply of liquid whey, rich in multiplied enzymes as well as assorted vitamins and minerals and milk proteins.

*Enzyme Booster:* Drink one or two glasses of this fermented whey in the morning, *before* breakfast, to help supercharge your sleepy enzymes with a vigor that will make you look and feel young for much of the day. You may add a squeeze of lemon juice for a tart taste if desired. This type of fermented milk is enjoyed by many of the long-living peoples of the world such as those of the Caucasus, Balkans, Middle East, India and Hunza, and the Sikhs. They have all discovered that whey is a miracle food with youth-building properties. It is an enzyme food that helps build and rebuild your digestive-intestinal tract. It is all-natural!

5. *Yogurt.* This ancient fermented food has long held the fascination of modern people who look upon it as a miracle for long life. Indeed, ancients reportedly reached the century mark and even beyond, enjoying youthful health attributed to this fermented food. Yogurt is fermented milk that is a prime source of enzymes, particularly acid-producing enzymes which create a healthful digestive-intestinal environment. This fermented milk food multiplies the enzymes which help to digest solids ingested in a normal meal. Yogurt enzymes convert milk lactose into lactic acid which supplements the action of the stomach's hydrochloric acid, promoting youthful, natural and drugless digestion!

Yogurt is especially beneficial for older persons. Its powerhouse of enzymes helps boost their own sluggish digestive juices to improve assimilation of nutrients. This is a valuable key to a feeling of youthfulness. When your digestive enzymes are boosted, you are able to better assimilate nutrients which helps promote better health and youthfulness. Fermented milk, such as yogurt, helps create this valuable rhythm. Yogurt enzymes inhibit the growth of the bacteria of putrefaction. Yogurt enzymes help to create an acid environment which helps the body absorb such valuable minerals as calcium, phosphorus and iron. This fermented food is a healthful enzyme tonic to be enjoyed daily.

*How to Make Yogurt:* Mix enough nonfat dry milk powder to make 1 quart of skim milk in 3 cups of water. Mix until the powder dissolves. Now bring to a slow boil over low heat. Remove. Cool to lukewarm. Pour ½ cup of the cooled skim milk into ¼ cup yogurt in a bowl. Stir well. Now add this mixture to the remaining skim milk. Blend thoroughly. Cover jar and place in a warm spot away from drafts for 5 to 6 hours. Refrigerate.

*Enzyme Booster:* Flavor with cinnamon, vanilla or honey, desired fresh fruits, or "spike" with chopped chives. Eat a cup at every meal. The benefit here is the yogurt enzymes are especially active at temperatures between 90°F and 118°F. In the digestive-intestinal system at these temperatures, the yogurt enzymes help boost digestive powers. Many persons who might otherwise have poor digestion will find that yogurt enzymes help strengthen the forces within the body to promote better assimilation of ingested foods. To finish a meal with a yogurt cup is an excellent way to send a shower of digestive enzymes into the system to help assimilate foods that have just been eaten. It's the natural, fermented way to help your system use nutrients in foods to build better health and youthful vitality.

**Use Fermented Foods Daily.** To create a healthful enzyme foundation, use fermented foods daily. This was one of the first methods discovered for the preservation of foods. Ancients knew what modern food scientists have corroborated–that the fermentation process acts to produce a natural lactic acid which boosts the action of youth-building enzymes.

**How Julia T. Helped Reverse Aging Clock with Magic Enzymes.** As a saleswoman, Julia T. had little time to prepare nourishing meals. She was always "on the go" with the result that she soon felt weak, tired and nervous. In particular, her face looked sallow, her eyes were bloodshot, her skin was dry. Although she was 30, she looked 50. She took time out to eat more wholesome foods but she was still wan, pale and listless. Her work suffered. Her memory was poor. Long columns of figures became blurred. When she faced the threat of losing her job, she sought help from a vivacious friend of hers. The secret was not so much in eating wholesome foods, the friend said, but in knowing how to assimilate the enzymes which would help make and remake body cells and tissues.

The friend told Julia T. to eat a raw vegetable or fruit salad before each meal. But more important, Julia T. was to have one fermented food between each meal. The benefit here is that the enzymes from the fermented food had a "double-action" power and would prepare the digestive system for the meal ahead.

**Simple Youth Restoration Program.** Julia T. had a glass of buttermilk before a whole grain and fruit breakfast. In mid-morning, she had a cup of yogurt. After lunch she would have a glass of sauerkraut juice. After dinner she would have a glass of whey. Whenever she wanted to munch on something, she would have the raw fermented cabbage. She emphasized wholesome raw foods for her meals. It took two weeks of a program calling for natural foods and fermented enzyme boosters and she soon responded favorably. Her skin was bright, her eyes were clear, her memory was better, and she worked with youthfulness. Her response was largely due to the fermented foods which enabled her enzymes to *assimilate* the youth-building nutrients of natural foods that she ate. A simple program, but it put her back on the road to restored youth!

**Enzymes: Small but Powerful Youth Activators.** Every living organism has its enzyme chains. Enzymes are needed to digest food to make it small enough to pass through the tiny intestinal pores directly into the bloodstream. Enzymes further help transport food to the muscles, nerves, bones, and glands. They help store excess food in the liver or muscles for future use as a natural "youth bank" right in your body. Enzymes help in the process of elimination of wastes and act as a self-scrubbing system. They enter into all known (and many unknown) body processes that help maintain good health and prolonged youth. Enzymes form a chain network within the body to regenerate cells and tissues and glands and organs to extend the life-youth cycle. They are magic workers that hold the key to a "forever young and healthy" body and mind.

*Summary:*

1. Enzymes are protein-like substances that activate the entire body's processes to help perpetuate the look and feel of youthful health.

2. Enzymes enter into almost every youth-health process in the body. Each of the some 700 different types of enzymes (and 100,000 enzyme systems) must function if body balance is to be enjoyed. There are 12 foundation "magic enzymes" that are stars in the entire network in your body.

3. Enzymes are found in natural, raw, nonprocessed foods. There are five "everyday foods" that help give you

abundant enzymes for youthful health. Use these fermented foods daily for a treasure of enzymes.

4. Julia T. helped reverse the aging clock with raw foods and special "everyday foods" that can be made in your home or purchased at almost any local produce market or health food store.

# 2

## *Five Fruits That Create "Fountain of Youth" Enzymes*

Fruits that are fully vine-ripened are Nature's most potent sources of youth-building enzymes. The fruit plant "eats" the rich store of nutrients from the soil, "drinks" the mineral-rich rains, "digests" their elements to create living juices within the emerging bud and flower. As the fruit emerges, it is warmed by the nourishing rays of the sun, oxygenated by the surrounding air, and ripened by Nature. This process creates an end product of youthful enzymes which are available for eating and nourishing by man.

Vine-ripened fruits are valuable sources of these youthful enzymes with few comparisons. The fruit plant is superior to the animal by its ability to take the natural elements of earth, air, water and heat and create a superior supply of amino acids, carbon chains, minerals, vitamins, a network of enzymes for its own self-nourishing food supply. The fruit plant can sustain itself because of its ability to manufacture its own youth-perpetuating enzymes and other nutrients. In contrast, the animal is unable to make its own complete supply of nutrients and must depend upon the plant for sustenance. There is little that the animal has acquired (except oxygen and water) that does not come from the plant. Nature has decreed that plants feed upon soil, and animals must then feed upon the products of plants. The embryonic plant holds the key to life on earth for all living things, including itself. It holds the secret to perpetual youth with its living enzyme supply that serves to give it life and to give life to others.

### *Why Vine-Ripened Fruits Are Sources of Youth-Building Enzymes*

Fruits are the edible parts of plants that result from the development of pollinated flowers. The soft, delicious pulp of vine-ripened fruits such as apricots, avocados, melons, papayas and pineapples are prime sources of enzymes. The seeds of such fruits are live, embryonic treasures, sending forth those ingredients that nurture and increase the enzyme content of the plant. When these fruits are allowed to fully ripen on the vine, their enzymes are at their peak power of sustenance. These enzymes can be seen at work as they nourish, sustain and regenerate the fruit which blossoms and glows with living youth and health. To the plant, its enzymes are like miniature "fountains of youth" from within, giving it youthful health. Vine-ripened fruits are prime sources of these elements which can then be introduced into the body to produce healthful properties. To obtain the most benefit from the enzymes in fruits, they should be organic (chemicals and pesticides destroy enzymes) and fully vine-ripened. Wash and/or peel them, and then eat as a fruit salad or just as a fruit in itself. The luscious taste, the enchanting aroma, the creamy smooth goodness of chewing, the luxurious blend of rare flavors offer more than just Nature's gourmet taste. Eating vine-ripened fruits is Nature's way of giving your body its own created rich treasure of youth-building enzymes.

While all vine-ripened fruits are nutritional troves of treasured enzymes, there are five such fruits that have distinct and special benefits. They should be part of your regular eating program in your quest for better youth and health with enzymes.

### *How Apricot Enzymes Rejuvenate the Eyes, Skin, Hair and Body*

The fully vine-ripened golden apricot is a treasure store of enzymes that work to send a stream of nutrients throughout the body that then help build and rebuild the health of the eyes, skin, hair and membrane networks that line the organism.

**Special Youth-Building Benefit.** Apricots contain the precursor carotene which is transposed into Vitamin A by enzymes. Enzymes take this vitamin and use it for rebuilding the delicate membrane linings of the eyes, the subcutaneous tissues of the skin, and the shaft of the hair. Enzymes work to help distribute Vitamin A throughout the body to help in the functioning of the thyroid and liver. These same apricot enzymes use Vitamin A with other nutrients in the vine-ripened fruit to stimulate the endocrine glands to send forth natural hormones that help guard against premature aging and help keep the body feeling healthy and looking young. Enzymes help take out the valuable nutrients from within the plant fibers to make them readily available for body use.

Apricot enzymes need the plant nutrients to help build skin and hair follicle health from within. The enzymes help to replace the decaying cells in the lower layers of the skin; they help "wash" the skin's oil sacs and pores. They further help guard against excessive dryness by making available a good supply of vitamins and minerals beneath the skin surface so that natural oils may then moisten and lubricate the "inside skin" to help keep the mucous membranes in a healthfully unctuous condition. Apricot enzymes are needed to send forth its own apricot nutrients to perform this vital youth-building benefit.

**How Arnold A. Helped Restore Youthful Vitality with Apricot Enzymes.** Bookkeeper Arnold A. was in his late thirties and thought that his blurred eyesight, dry and flaky skin, and constant sniffling could all be traced to his indoor confinement and long hours of poring over accounting books. When his eyes became red-rimmed and he caught one cold after another, when Arnold A. began walking with a stooped gait and looked and felt older than his age, he decided that something had to be done.

**Simple Change Begins Enzyme Recovery.** Arnold A. believed in eating fruit, but his error was in eating cooked, canned, stewed or processed fruit. The enzymes were destroyed in the heating and manufacturing. He was told that raw, rich, uncooked fruit, especially raw golden apricots, were long considered to be treasures of youth-building elements. Arnold A. eliminated all cooked fruits. He substituted with natural, fresh raw fruits. He ate fresh vine-ripened apricots almost daily. This simple change produced wonders. The enzymes in the raw vine-ripened apricots worked to activate Vitamin A in his system to then become transformed in a usable form. (Without enzymes, Vitamin A may remain inert and useless.) Enzymes took the Vitamin A to help nourish the cellular membranes around the eye, then to nourish the skin cells, lubricate the hair follicles, nourish the endocrine gland system, and promote a feeling of well-being. Raw fruits, especially apricots, helped Arnold enjoy a clearer skin, resistance to cold-catching and respiratory ailments thanks to an enzyme-lubricated respiratory tract, as well as better hair health with natural lubrication. In particular, the apricot enzymes helped improve his eyesight, which became clear, bright and youthful. Arnold A. now looks and feels much younger thanks to the all-natural program of fresh raw fruits eaten daily, and the regular use of vine-ripened apricots with their youth-building enzymes.

### *How Avocado Enzymes Help "Wash" Body Arteries*

This romantic fruit was worshipped by the ancient Aztecs who reportedly regarded it as the source of eternal youth. The long-living healthy Aztecs ate avocados regularly and enjoyed smooth healthy skin, flexible limbs and youthful agility even at an advanced age. They may not have identified the "secret" of the youth-building power of the avocado, but they knew that it held "magic" properties that made them feel young and supple. Legend has it that warriors ate avocados daily so their arms and legs could move with youthful flexibility.

Today we know that the avocado contains valuable enzymes that take up its unique supply of essential fatty acids for use in body lubrication.

**Special Benefit of Avocado Enzymes.** This vine-ripened fruit contains unique enzymes that are able to take up the plant's supply of three types of essential fatty acids (linoleic, linolenic and arachidonic) that are helpful in keeping the body's arteries and veins in a "washed" or lubricated youthful condition. Enzymes take these essential fatty acids and use them to "scrub" the arteries, helping to promote more youthful flexibility. The Aztecs recognized this magic power of avocados and emphasized their use in daily eating programs. Their superior health is evidence of their awareness of the benefits of eating avocados for youthful vitality.

Enzymes in the avocado will use the essential unsaturated fatty acids to help bring about a significant reduction of the serum cholesterol in the bloodstream. These enzymes will take up the 11 vitamins and 14 minerals of the avocado to help nourish the body from head to toe. Enzymes will also take up the plant protein in the avocado and use this precious form of growth-substance for the building and rebuilding of the body. Virtually no other plant has such a supply of protein as the avocado. Nature has placed an abundance of enzymes in the avocado that take up this plant protein and transform it into usable amino acids for use in enriching the bloodstream, in helping to guard against cholesterol buildup, and in helping to moisten and lubricate body channels and networks to keep them supple and flexible. Avocado enzymes eagerly take all the valuable substances in this vine-ripened plant and assist in metabolism in the body to help extend the prime of life. Small wonder that the Aztecs worshipped the avocado as the "plant of eternal youth."

**Secret Artery-Washing Power of Avocado Enzymes.** Avocado enzymes take *phytosterols* from the plant's tissues, and use these fat-dissolving substances to help "wash" the arteries and help keep down the serum cholesterol levels. *Phytosterols* found in the avocado plant are natural "scrubbers" used by the enzymes to help keep the arteries and nerve networks in clean condition.

**How a Telephone Operator "Loosened" Tight Fingers and Joints with Avocado Enzymes.** Mary C. found it difficult to work her switchboard with her gnarled fingers. Holding a pencil became painful when her joints just "refused" to unbend at will. Mary C. found that her body seemed "tight" and her flexibility was poor. She brooded over the fear of developing arthritic-like stiffness. It would interfere with her work as a telephone operator that demanded she move easily with youthful hands, wrists and arms. She felt distressed.

**How "Avocado Enzyme Fats" Helped Control Cholesterol.** Mary C. liked fats in her foods. No doubt her high intake of meat and butter fats gave her an increase in the serum cholesterol rate that threatened a predisposition to arteriosclerosis and arthritic-like stiffness of her arms. How to keep enjoying all the "rich flavor of fat" but with less of the risk of cholesterol deposits? Simple. Mary C. was advised to eat avocados in place of other fats such as butter or lard. She was told to trim off all visible fat from meats. Furthermore, she was to use mashed avocado pulp as a substitute for butter and other "hard" fat. She was to satisfy her fat "urge" with a vegetable fat found in the avocado. But it was more than just the substitution that eventually helped loosen up her gnarled fingers and joints. It was the natural work of avocado enzymes that began by helping to slough off the accumulated fats of her arterio-neural network. Then the avocado enzymes took up the plant's own supply of unsaturated fatty acids and used these to help "loosen" and wash away the accumulated sludge of the hard fats.

By cutting down on hard fats and substituting avocado fats, Mary C. enabled avocado enzymes to transport the plant's essential fatty acids to the arterial network and help promote a feeling of lubrication. Now Mary C. had flexible fingers and joints. She worked again with the agility of youth.

A unique benefit here is that the vine-ripened avocado contains *no starch* and *almost no sugar.* This means that its plant enzymes can work *without interference* from either starch or sugar, and are unhampered by starch-sugar metabolization. These plant enzymes can work full force upon the avocado's store of essential unsaturated fatty acids to extract them from the pulp and send them directly into the body for artery "washing" and youthful resiliency of the arms, legs and joints. The avocado is the Aztec symbol of perpetual youth, thanks to its rich supply of enzymes and essential fatty acids.

### *How Melon Enzymes Help Produce "Instant Energy"*

When any kind of a melon is permitted to fully ripen on the vine, it is a prime source of Nature-created enzymes that help produce "instant energy."

During the ripening process, the fructose content of the melon increases. This gives the vine-ripened melon the naturally sweet and juicy taste. Fully vine-ripened melons are top-notch sources of fructose which is brimming with enzymes that use this natural fruit sugar to create a form of energy that may rival that of a youngster.

*Special Benefit:* Enzymes in fructose found in vine-ripened melons are able to immediately send this energy supply directly into the bloodstream. Fructose enzymes that are created by the full ripening process are in a form that can be used immediately by the bloodstream. Once fructose enzymes enter the bloodstream, they help create a healthful blood sugar level to provide a steady boost of energy.

*Blood Enrichment Benefit:* Melon enzymes also take up the plant's supply of Vitamin A to promote a *hematogenic* benefit; that is, they help enrich the blood, cleanse it of accumulated impurities, act as a natural disinfectant, and then take up the fructose and send it to the liver to be stored and used as needed. It is this double-action benefit that helps boost energy almost immediately. The stored fructose is taken out of the liver by the enzymes and sent via the enzyme-cleansed bloodstream to the cells and tissues of the glands where a natural and youthful energy boost is created. Enzymes therefore cleanse the bloodstream so that it can speedily transform the stored up supply of melon-created fructose to body glands to promote youthful vitality. It is a miracle of Nature that adds up to better health for you.

*Note:* It is vital to obtain fully vine-ripened melons as well as all other fruits. The enzyme content of any fruit is at its peak level when that fruit is permitted to fully ripen on the vine. To obtain top quality enzyme-propelled fructose that goes to speedy work in boosting the hypoglycemic levels in the bloodstream, use melons (and other fruits, too) that have been allowed to fully ripen on the vine.

**Five Melons with High Enzyme Values.** While most vine-ripened melons have high enzyme values, these five varieties appear to offer a top quality level of enzyme with good fructose levels to create a healthful balance:

1. *Cantaloupes* are good sources of vitamins, minerals, and natural fruit sugars as well as youth-building enzymes that take these nutrients out of the pulp and put them to speedy assimilation in the bloodstream. *Purchase Tip:* At the stem end, where the fruit joined the vine, there is a slightly sunken scar. When the scar is smooth, it indicates the melon

was ready to "slip" or separate naturally from the vine at harvest time. This is a prime source of enzymes.

2. *Casaba melons* are globular in shape, pointed at the stem end. *Purchase Tip:* Fully vine-ripened melons have a buttery yellow rind with wrinkles and lengthwise furrows. The fruit pulp is brimming with enzymes if it is creamy white and meltingly juicy.

3. *Crenshaw melons* are globe shaped with a pointed and slightly wrinkled stem end. If the rind is green, the melon is not yet ripened. *Purchase Tip:* When the Crenshaw matures, the skin color becomes mottled with yellow; it turns to an almost solid yellow-gold color at the peak of flavor and ripeness. The fruit pulp is orange-yellow and thick with a slightly spicy aroma that is indicative of its powerful enzyme maturity.

4. *Honeydew melons* are bluntly oval in shape with a smooth rind. Often this type of melon is cut from the vine before reaching the fully ripe stage and it may be left at room temperature to continue the enzyme ripening. *Purchase Tip:* When the rind changes from a slick whitish green to a creamy white and feels slightly tacky to the touch, then its enzyme power is at a high peak. The fruit pulp is a delicate green, juicy, very sweet, fine grained and almost melting in texture. Its delicate, sweet, pleasant aroma indicates its rich enzyme supply.

5. *Persian melons* resemble cantaloupes. The surface is more webbed than netlike. The background is grayish green. *Purchase Tip:* When the background is brownish and fruit pulp is thick and varies in color from pale orange to salmon pink, then it has a peak enzyme content. The texture should be finer and firmer than the cantaloupe but with its own distinct and juicy good flavor.

*About Watermelons:* These have smaller amounts of enzymes because they are mostly water with goodly amounts of vitamins and minerals. Watermelons have "milder" enzymes because the natural water content has made them diluted. It is wise to eat watermelon together with other vine-ripened fruits to maintain a good balance of enzymes as well as other nutrients. *Purchase Tip:*

Watermelons should be symmetrical in shape, with a good color and a fresh bloom to the surface. The underside spot where the watermelon touched the ground should have a yellowish tinge that indicates vine-ripening with goodly amounts of enzymes and other Nature-created health-building nutrients.

**How a Cab Driver Uses Melon Enzymes for Early Morning "Instant Energy."** Larry J. needs early morning "get up and go" to keep alert on the roads as a cab driver. He knew the values of fruit and would eat a large fruit salad together with a popular packaged breakfast cereal in the morning. But Larry J. found himself getting drowsy in an hour or so. The promised "energy" from fruits seemed false to him. When he dozed off in front of the wheel and almost had an accident, he decided to do something to keep himself alert and awake in the morning.

**Simple But Powerful Adjustment Gives Him Energy.** Larry J. was told that he was eating the wrong fruit and wrong cereal. He was eating *canned* fruit and eating *processed-devitalized* packaged breakfast cereals. The canning and processing methods used heat, dehydration and chemical preservatives. These methods destroyed and "killed" the perishable enzymes. So Larry J. was eating *enzyme-dead foods.* He had to make a simple adjustment. Here it is:

*Enzyme Energy Booster:* Each morning, Larry J. would eat an assortment of fresh melon fruit slices. The melons would be vine-ripened. He would sprinkle them with some raw wheat germ for added enzyme-boosting powers. He would add a spoonful of natural honey to provide the enzymes with predigested sugars that would then propel the enzymes from the fruits into his bloodstream to give him the coveted energy. Larry J. switched from lifeless processed breakfast cereals to a natural whole grain cereal. Either he had granola, Swiss-style muesli (available at many large supermarkets and most health stores) or whole grain oatmeal, millet, whole wheat cereal or any other type of unbleached, nonprocessed natural grain cereal. The readily available supply of natural grain carbohydrates acted as a "lightning spark" to the enzymes from the raw fruit, and together with the "boost" from the natural sugar in honey, it added up to an Enzyme Energy Booster that gave this cab driver the vigor and vitality of a very young person! Now Larry J. could compete with younger cab drivers for vitality, energy and "get up and go." Most important, this simple melon and grain breakfast gave him "instant power" when he needed it most–in the morning–and then some!

### *How Papaya Enzymes Create "Young Digestion"*

Called a tree melon because of its resemblance, this legendary fruit grows on a giant herbaceous plant. It is a tropical fruit. During the process of vine-ripening, the papaya is warmed and nurtured by the sunlight. The ultraviolet rays penetrate the rind and set off a working factory of nutrients that labor to create a powerful enzyme known as papain. From time immemorial, the fresh leaves of the papaya plant have been used by the Indians to wrap meat in. The enzymes from the leaves are able to make meat tender. Papaya, the "wonder enzyme" fruit is reported to be richer in enzymes and vitamins than most any other fruit.

*Unique Benefit:* The papaya has almost no starch, either in the green or ripe fruit. The ripe pulp has a powerful supply of enzymes that are not hampered by the presence of starch. This means that papaya enzymes with the reduced sugar form offer a bonus of instant vitality in the bloodstream. The simple forms of sugar are taken by the papaya enzymes and then sent into the digestive system via the bloodstream.

**Creates "Young Digestive Powers."** During the vine-ripening process, most of the elements have undergone predigestion by the enzymes in the fruit. The papaya's active enzyme principle helps to digest foods in the stomach. It works speedily because papain can be retained in the digestive system as an always-available enzyme to help boost digestive functions.

**Nature's Morning Tonic.** The juice of this vine-ripened fruit helps to correct "morning congestion" *if taken on an empty stomach.* The benefit here is that papaya enzymes are propelled by the predigested fruit sugars to create a pronounced tonic and rebuilding effect upon the digestive system. Because papaya juice is unhampered by starch, it is especially quick acting in the morning. It offers instant relief and helps cleanse the stomach so it will function better for the day ahead.

**Valuable Nutrients.** Fresh papaya pulp also contains some amounts of protein, natural citric, malic and tartaric acids, several minerals and an abundance of vitamins. All are "sparked" by papain, the enzyme in the fruit. This helps create a form of internal rebuilding that adds up to a youthful digestion.

**How Papaya Enzymes Help Rejuvenate Digestive System.** Papaya enzymes offer soothing relief for problems of gastritis, dyspepsia and indigestion. The enzymes attack and destroy dead tissue and decaying membranes in the digestive tract. The enzymes then help to cleanse the intestinal tract of accumulated decayed and infectious matter. They also help promote more normal body functions. Papaya enzymes in the digestive system will help assimilate ingested foods. Papaya takes out the nutrients from these foods and puts them to work in overall head-to-toe assimilation via the bloodstream. Papaya enzymes also create a beneficial acid-alkaline environment. They help form amino acids and have shown some beneficial action by digesting "tough" foods. Papaya enzymes are often considered the "antidote" for almost any form of indigestion. They are a miracle of Nature.

**How Jean L. Remade Her Digestive System with Papaya Enzymes.** Jean L., a nervous housewife, had chronic indigestion and embarrassing constipation that could swerve into colitis. Whatever she ate felt like a leaden weight in her stomach. She lost weight, became flabby, showed memory impairment, and became chronically fatigued. (*Note:* The secret of youthful health is in proper assimilation! Like many others, Jean L. may be eating wholesome food but it is not being assimilated and the available nutrients are passed off or improperly oxidated or metabolized in the system. The key to youthful health is in youthful digestion of wholesome foods. Without such digestion, top quality food is wasted!)

**Tropical Gift Opens Doorway to Health.** A friend vacationing in the tropics mailed Jean L. a box of vine-ripened papayas. Curious about the golden-green fruit, the housewife ate several of them. She discovered a wondrous feeling of internal regularity almost from the beginning. The magic power of papaya enzymes is in its "instant" ability to create healing virtues. Jean L. enjoyed freedom from her indigestion. Bowel regularity was restored. Now she began to fill out. Her skin was firm. Even her memory improved. Her energy was on the rebound. The secret here is that the papaya enzymes helped assimilate foods. The vitamins, minerals, essential fatty acids, natural sugars, and amino acids, were seized up by the enzymes and poured into the bloodstream where they went to speedy work to open up a doorway of health for the tired housewife. Now Jean L. could enjoy life again. Her digestion became so improved that she ate with the gusto and assimilation powers of her teenage daughter. She eats one fully vine-ripened papaya a day as her "natural medicine" and enjoys "forever young" digestion. She looks it too!

### *How Pineapple Enzymes Reduce Swelling and Inflammation*

The thorny vine-ripened pineapple is a graceful fruit that has often served as a table decoration and has always been regarded a symbol of hospitality. Today the pineapple has been found to contain more virtues. This luscious fruit is brimming with enzymes that reportedly are able to soothe swelling and inflammation whether from a severe bruise or conditions of rheumatoid arthritis.

*Special Tip:* The best source of enzymes is the pineapple that has been fully ripened on the vine. If it is picked green, it means that its enzyme and vitamin content is very low. When you purchase a pineapple, it should have a fresh, clean appearance. The color should be dark orange-yellow and the fragrance should be pleasant. The heavier a pineapple, the higher is the quality of its enzyme content. The "eyes" should be flat, almost hollow. This is Nature's way of telling you that the pineapple is at its enzyme peak.

**Fresh vs. Canned Pineapple.** The fresh, nonprocess pineapple is your prime source of protein-digesting enzymes. The canned pineapple is lower in vitamin content (the B complex and C vitamins are water soluble and lose potency during processing; these vitamins are needed by enzymes for maximum health rebuilding). Furthermore, canned pineapple will have a lower and weaker potency of enzymes, if they survive the processing. You would do well to obtain fresh pineapple. Its thick skin protects it against poisonous insecticides that seep through the skins of other fruits. When you peel off this thick, thorny pineapple skin, you are getting down to the pure fresh pineapple pulp, rich in natural vitamins, minerals and enzymes. Pineapple enzymes work swiftly because a normally and fully vine-ripened pineapple is about 13 percent sugar which is used by the enzymes for speed and efficiency.

**How Pineapple Enzymes Promote Healing.** Among the many enzymes in vine-ripened pineapple is the *bromelain* group. Bromelain enzymes reportedly protect against inflammation. They guard against edema (excessive liquid accumulation) in regions injured by infections, wounds or illness. Bromelains are also protein-digesting enzymes which makes them miracle healing agents.

**Magic Power of Pineapple Enzymes.** Bromelains are able to help heal inflammation and edema. During the inflammatory process, fibrogen (a soluble protein in blood plasma) is converted to a soft, partially polymerized insoluble fibrin (another protein) in the walls and tissues of small blood vessles. The word "polymerized" refers to the clogging together of small fibrin molecules to create much larger and more complex molecules. The problem here is that deposits of partially polymerized fibrin inhibit the permeability of tissues and vessels. These deposits block the return of edema fluid into the circulation. The use of pineapple enzymes helps to "unblock" these deposits.

**A Doctor Reports on Healing with Pineapple Enzymes.** During a five-year period at Philadelphia's Municipal Welfare Clinic (a health facility for government policemen, and firemen and their families), M.G. Cirelli, M.D., treated some 700 patients who had problems of inflammation, edema traced to problems of bruises, contusions, abrasions, sprains, wounds from surgery, boils, cellulitis (spreading inflammation of connective tissue) and varying forms of ulcers. Dr. Cirelli treated them with simple bromelains—enzymes found in pineapples. He reported[1] that "The bromelains may act by depolymerizing these occlusive deposits, thereby restoring draining and increasing permeability." He feels that bromelains do not interfere with the normal process of fibrin which acts as a life-saving blood-clotting mechanism. Dr. Cirelli adds, "Therapy with bromelains for the condition indicated was usually followed by a marked subsidence of pain, edema and inflammation. The increased mobility and decreased period of disability facilitated a quicker return to normal activities. The time interval between admission and discharge of patients treated with

1. "Bromelains for Healing," *Clinical Medicine* (June 1967), pp. 168-172.

bromelains was estimated to have been reduced by 30 to 50 percent. No toxic effects attributable to the medication were observed."

**Natural Bromelains are Safe.** All researchers reported that these natural pineapple enzymes promoted effective healing without any side effects!

**Recoveries are High.** Dr. Cirelli notes that 62.5 percent of his patients made an "accelerated recovery." There were other benefits from this bromelain program. The use of these natural pineapple enzymes made it unnecessary to use the conventional (and disliked) methods of splints, bandages, heat, compresses, and immobilizing devices.

**Recovers in 48 Hours.** One 50-year-old man who had scrotal cellulitis with edema was speedily healed. "After 48 hours of therapy with the bromelains," writes Dr. Cirelli, "2 tablets, 4 times daily, the swelling was reduced by half. Four days later the condition was completely cleared and the patient was discharged." Bromelain enzymes are also available in tablets at most health stores or pharmacies.

**Healed of Ulcers.** One 33-year-old woman had a painful buttock ulcer caused by an X-ray burn received a few years previous. The woman said that skin grafts and "all kinds" of medication did her little good. She was given four months of bromelain enzyme treatment. The entire ulcerous region was healed and she was discharged in good health, thanks to this pineapple enzyme.

**Enzymes Heal Fracture.** A fireman had a fractured wrist. The region was swollen, and there was pressure on his cast. (This impairs blood circulation and if uncorrected, can lead to blood poisoning and possible loss of the limb.) The clinic doctors put the fireman on bromelain enzyme treatment. Within 24 hours the pressure and the edema were gone. The fireman's fingers felt warm. He could move them freely, showing healthy circulation. The swelling had gone down so much that he had lots of room in the cast. The enzymes had helped heal the symptoms which complicated the fracture.

**Helps Heal Thrombophlebitis.** This type of blood clot and venous inflammation can respond to pineapple enzyme therapy. In a special test,[2] Bert Seligman, M.D., treated 73 patients hospitalized for acute thrombophlebitis. All but 27 were given bromelains in a special test. Day after day, doctors tested these bromelain-

2. "Bromelains for Thrombophlebitis," *Angiology* (January 1969), pp. 240-244.

receiving patients and reported there was improvement in the degree of pain, edema, redness, tenderness, elevated skin temperature and disability. The non-bromelain patients had less improvement under standard medication.

**Helps Soothe Rheumatoid Arthritis.** These pineapple enzymes help heal the "leftover" soft tissue inflammation often accompanying joint involvement in rheumatoid arthritis. Two physicians, Abraham Cohen, M.D., and Joel Goldman, M.D.,[3] treated 29 patients with problems of rheumatoid arthritis, osteoarthritis and gout. They gave their patients a goodly amount of bromelain enzymes and reported excellent or good results in 72.4 percent of the cases. The doctors explain that bromelain enzymes are helpful in bringing down inflammation. Once the joint becomes inflamed it is no longer active. This leads to soft tissue swelling. Fluid becomes "trapped" in the adjoining tissue. Pineapple enzymes help by depolymerizing fibrin deposits and helping to restore drainage.

**Enzymes Reduce Swelling.** A 30-year-old woman was examined for quiescent rheumatoid arthritis and a residual swelling around her right knee. Drs. Cohen and Goldman gave her pineapple enzymes and the swelling was soon reduced appreciably. "There was a return of some of the swelling when bromelains were stopped, which disappeared when bromelains were again administered," say the doctors.

**Enzymes Restore Finger Flexibility.** A 34-year-old woman had a problem of swollen joints of her fingers and wrist. The doctors treated her with steroid drugs but there was no reduction in swelling. The fingers and wrist were inflexible. The doctors then gave her bromelains and now the swelling was reduced and finger flexibility restored.

**How Pineapple Enzymes Improve Digestion.** Many persons lament the passing of adolescence when digestive powers were strong. Yet the use of pineapple bromelain enzymes can help give an older person the digestive power of a youngster. Enzymes are proteins. They are digested in the intestinal tract into their amino acid components and then are used to help metabolize foods. Better assimilation is possible with pineapple enzymes.

Joseph Miller, M.D.,[4] put a set of patients on three days of bromelain therapy. "An increase in the amount of digestion was

3. "Therapy for Arthritis," *Pennsylvania Medical Journal* (June 1964), pp. 212-217.
4. "Enzymes and Inflammation," *Experimental Medicine and Surgery*, XXII, (1964), pp. 293-299.

found in every case. These tests provided evidence that the enzyme had been absorbed from the small intestine." Furthermore, bromelains are able to give a boost to the body's own built-in natural defenses against illness because of this better assimilation. Illness or trauma (stress, strain, shock) causes an increase in proteolytic (protein-digesting) activity in the serum (clear yellow fluid that separates from clotted blood). Bromelains promote better absorption and assimilation of body processes and this helps muster body forces to resist effects of illness. Undoubtedly, recovery is higher too.

**Pineapple Enzymes Speed Healing.** When the body has an injury or infection, proteolytic agents are swiftly drawn out of the bloodstream to the specific site of injury. This may leave the bloodstream more vulnerable to other infections. A large supply of bromelains in the bloodstream means that these enzymes will be available to be sent to the infection site to promote the marvel of healing. Furthermore, the presence and availability of pineapple enzymes for healing, means that the bloodstream will not have to be depleted of much of its proteolytic agents which are needed for infection-fighting elsewhere in the body. Bromelains, pineapple enzymes, are Nature's own disease-fighters. They promote disinfection, ease swelling, create youthful digestion and help you bounce back to good health with speedier "youthful" recovery rates. From head to toe, and from the inside and the outside, pineapple enzymes promote youthful healing.

Vine-ripened fruits are Nature's prime sources of all-natural "fountain of youth" enzymes. Use them to tap this source of youthful health.

*Highlights:*

1. Apricot enzymes help rejuvenate eyes, skin, hair and body. Arnold A. received youthful vitality with simple apricot enzymes.

2. Avocado enzymes help "wash" body arteries with a rich supply of essential fatty acids and interdependent vitamins, minerals and unique plant proteins. Mary C. used this Aztec fruit to improve her health and loosen up tight fingers and joints.

3. Melon enzymes combine with natural fructose to create a rush of energy via the bloodstream. Cab driver Larry J. was a young man again behind the wheel, thanks to melon enzymes in a delicious and energy boosting enzyme breakfast.

4. Papaya enzymes help create young digestion and better assimilation. This tropical fruit helped Jean L. remake her digestive system with youthful gusto.

5. Pineapple enzymes reportedly helped reduce swelling, inflammation, ulcers, thrombophlebitis and forms of rheumatoid arthritis. They act as natural "blood boosters" to resist ravages of daily living and to help ease symptoms of shock and strain.

# 3

## *How Natural Fruit-Vegetable-Grain Enzymes Help Promote a "Forever Young" Skin*

Raw fruits, vegetables and grains are prime sources of tiny "wonder workers" that are able to rejuvenate the millions of skin tissues and cells that are constantly in need of repair and replenishment. These rejuvenating enzymes, when applied to the *outside* of the skin, are able to create a "renewal" activity to slough off age-causing decay and waste. These enzymes are able to seep within the pores of the skin to help replenish nourishment from beneath the surface. Enzymes are natural "skin foods" in that they bring about a catalyst-like double-action to the tissues and cells to help them repair and regenerate themselves. The enzymes do not become part of the skin, but they promote the needed replenishment that helps keep the skin looking and feeling young. Enzymes help create the bloom of youth for your skin at all ages.

### *How Lemon Enzymes Help Restore Youthful Beauty to Your Skin*

**Weeps Because She Looks Older Than Her Husband.** Martha B. wept to overhear neighbors whisper that she looked older than her husband. She had tried a wide variety of commercial cosmetics but their benefits were only temporary. Without heavy makeup, her skin looked blotchy and had ugly exaggerated pores and unsightly "aging spots" that did not go away no matter how many home treatments she would give herself with an endless array of creams and lotions. Martha B. looked and felt much older than her husband, and it hurt her to hear the malicious whispers of neighbors. It was one kind neighbor who always had the "bloom of youth" on her skin who told Martha B. a "million-dollar youth secret" that cost "just pennies" and took "just a minute" in the privacy of the home. Here is this reported "million-dollar youth secret" that Martha B. used upon herself:

**Use Fresh Lemons for Youthful Skin.** Squeeze the juice of a fresh lemon into a cup. Now use the lemon juice for these rejuvenating benefits:

1. *Skin Freshener.* Dip your fingers into the cup of lemon juice and apply this astringent-like liquid to your skin. It leaves your skin feeling tingly clean and helps to remove excess dirt-laden oil too.

2. *Facial Mask.* Lightly beat an egg white. Mix in a tablespoon of mayonnaise and a tablespoon of lemon juice. Spread on face and neck and allow to dry. Let remain for one minute. Now rinse with warm and then cold water. *Secret Benefit:* Lemon enzymes help to draw out the impurities from the skin, tighten the pores and refine the complexion, giving it a youthful glow.

3. *Skin Rejuvenator.* Mix an egg white and half a cup of lemon juice with sufficient dry oatmeal to make a paste. Apply to face and neck and allow to dry. Let remain one minute (or longer, if preferred). Now rinse with warm and then cold water. *Secret Benefit:* The lemon enzymes use ingredients in the dry oatmeal to "scrub" the pores of the skin and to help lift out clogged impurities. When washed off, the skin has a fresh, lively and healthy look of the clean outdoors, thanks to enzymes in lemon juice.

4. *Soft Hands.* Rub the hands with lemon juice after dishwashing. *Secret Benefit:* Enzymes in lemon juice help create a synergistic action of the skin glands to send forth a supply of natural acid that guards against drying and chafing. The lemon enzymes restore this natural acid mantle to the skin; without this natural acid condition, bacteria can grow and cause premature aging and wrinkling of hands. Lemon enzymes create a restoration of this necessary age-fighting acid mantle of the skin of the hands and body too.

5. *Refresh Tired Feet.* A wedge of lemon rubbed on the feet provides refreshing tingly relief. Or soak the feet in comfortably hot water to which you have added the juice of one lemon. (It deodorizes too!) *Secret Benefit:* Lemon enzymes help restore cells and tissues to the worn and abused pavement-pounding portions of the feet. This is a natural renewal of needed skin cells that helps provide refreshing comfort as well as softness.

**Martha B. Helps Roll Back Years with Lemon Enzymes.** It took up to one week of daily lemon enzyme treatment at home before Martha B. began to look better. The lemon enzymes helped "skin off" and "peel away" decay and waste. Result? Martha B. now looked clean, fresh and so youthful that she was mistaken for her husband's younger sister. Thanks to lemon enzymes, her skin enjoyed a natural and healthy comeback!

### *How Citrus Fruit Enzymes Hold the Secret of Skin Youth*

The enzymes of any raw citrus fruit (orange, lemon, grapefruit, melon, berry, tangerine, nectarine, apricot, peach, plum, pear, etc.) hold the secret of skin youth by helping to restore the natural acid mantle to the surface.

**Normal Acid Condition Is Rejuvenating.** The skin needs a natural pH mantle that has a range of 4.5 to 5.5. In order for the skin to be healthy, smooth, soft, it has to be properly acid-balanced between those ranges.

**Soaps are Alkaline.** Most soaps used to wash the skin are highly alkaline. When you wash your skin, the natural pH (acid) mantle is removed. This leaves the skin vulnerable to the ravages of daily living which include pollution, soot, dirt, grime, wind, sand, heat and cold. Without this natural pH (acid) mantle, the skin is susceptible to aging. There are some soaps that claim to be acid-balanced but they are usually balanced at a pH of 6.5 to 7.5 or even as high as 8.5. This means that these soaps are as much alkaline as they are acid and their use is self-defeating. The key to a youthful skin is in helping to maintain a natural pH (acid) mantle of 4.5 to 5.5. Nature holds the key to helping to restore that balance.

**How Citrus Fruit Juices Help Keep Skin Soft and Young.** The fresh juices of any citrus fruit are prime sources of enzymes that are able to restore a natural acid covering. Fruit juices are naturally acid because the fruit plant contains enzymes which cause a "ferment" to create this acid level. When these enzyme-containing fruit juices are applied to the surface of the skin, they help to recondition the glands and tissues beneath the surface. They help to stimulate sluggish skin cells to emit their own supply of necessary acid. Citrus fruit juice enzymes help rejuvenate skin that may have become too alkaline through excessive soap washing.

**Enzyme Skin Food Makes Sophie Look "Too Young."** As a travel guide, Sophie D. was constantly exposed to the harsh elements of the outdoors. When indoors, she was barraged with clouds of tobacco smoke from others in the travel agency, not to mention the drying effects of either heating or air conditioning. Sophie washed with alkaline soap. The problem here is that the skin is about 85 percent protein that "feeds" upon its natural acid mantle. Sophie's washing removed the protective mantle, "starving" the protein. Combined with overexposure to the elements of the outdoors and indoors, her skin lost its moisture balance. It began to flake off in unsightly dryness. Lines marred the once-youthful skin. A traveller told her about using the juice of any fresh citrus fruit as a "wash" and as a way of "feeding" the skin.

Sophie D. would use orange juice as a "wash" after her regular soap and water treatment in the morning. At noontime she would use a grapefruit juice "wash," and at night she would squeeze the juice of a grapefruit in a tub of comfortably warm water and luxuriate for a half hour.

The benefit here is that the enzymes helped the skin replenish its own supply of needed cells and tissues. The raw juice enzymes then "fed" the skin, prompting it to break down proteins into small amino acid molecules that "plumped up" the texture and smoothed out furrows and creases. Sophie's skin looked fresh, glowing and healthfully clean. An envious airline stewardess said that Sophie D. looked "too young." It was the finest compliment anyone could give her. She felt a debt of gratitude to the traveller who had told her about citrus fruit enzymes. She was forever grateful to the enzymes themselves.

## How Cucumbers Help Create Natural Moisture for Dry Skin

The cool cucumber offers a treasure of moisturizing agents that help create natural skin smoothness. Enzymes in raw cucumber juice act as natural emollients by stimulating the skin's own sluggish cells to send forth a supply of needed moisture. Cucumber enzymes help feed the sebum glands beneath the skin surface and help soften the appearance of dry skin.

*Cucumber Lotion:* Peel a nice firm cucumber and cut into very thin slices. Now put the slices in a blender with very little water in order to get a soft, mushy pulp. (You can also do this by chopping well or mashing with a fork. Put the cucumber pulp into a cheesecloth square and press out the juice into a small bowl.) Now apply this cucumber juice as you would any skin freshener: pat it on your skin and let it dry on your face. If possible, let it remain most of the day. The benefit here is that the cucumber enzymes will go beneath the skin surface to help the sebum glands "drink" and then send out a natural supply of moisture to combat dryness–the bane of aging skin.

## Overnight Berry-Rose Enzyme Skin Pickup

Press out the juice of a cup of seasonal berries (strawberries are best) through a sieve into a bowl. You should have about a half cup of rich red berry juice. To this, add a half cup (4 ounces) of rose water, available at most pharmacies. Chill before use, or add an ice cube to the mixture before you pat it on your face. If possible, leave this Berry-Rose Enzyme Skin Pickup on your skin while you sleep.

*Enzyme Benefit:* The raw juice enzymes are perked up by the rose water to create a natural *humectant* action beneath the surfaces of the skin. The berry enzymes are water loving and can take moisture from oxygen and hold it to the skin, helping to make it look plumped out, soft and dewy. It's a natural way to help give your skin cells the needed moisture for better appearance and a feeling of smoothness.

### *How Plant Enzymes Help Cleanse Skin Impurities*

If you are troubled with blackheads, blotchy pores, dark lines and wrinkles, take advantage of the cleansing power of enzymes in a simple folk skin healer:

*Oatmeal-Avocado Masque:* Use a half cup of ordinary oatmeal, preferably *not* the instant kind. Add about a tablespoon of milk or hot water. When the oatmeal absorbs the liquid, combine it with the mashed pulp of half an avocado until they are well blended. Now, *before you apply,* wash your face thoroughly, splash on some lemon juice and let dry. Then apply this Oatmeal-Avocado Masque evenly. Relax while it dries for about 15 minutes. Let remain up to an hour or longer. Then sponge off with warm water, followed by a few clear rinsings. Finish with a fruit juice splash.

*Benefit:* The enzymes in the avocado take up the minute particles in the oatmeal to cause a "grain rub" that helps "lift out" skin impurities in the pores and creases of the skin. Simultaneously, the avocado enzymes work to lubricate the skin to bring a new glow to even the most lackluster complexion.

### *Hollywood Secret Face Scrub*

Many of Hollywood's most "eternally beautiful" theatrical stars reportedly paid a small fortune to receive this special treatment that helped give them young and glowing skins. Hollywood personalities must look fresh and glowing in order to compete with younger performers coming on to the scene. This secret face scrub reportedly worked miracles in moments for many aging stars. The secret filtered out and now you can use it yourself, for a few pennies a treatment, right in your own home.

*Secret Treatment:* Peel and mash a ripe avocado (one half should be enough), mix it with a half cup of finely ground yellow or white cornmeal. Before you apply, wash your face thoroughly and finish with a splash of raw fruit juice. Now gently rub in this avocado-cornmeal paste. Rub all over, especially around those trouble spots which have wrinkles, impurities, unsightly pores, big blotches, and blackheads. Keep rubbing. Let it remain about five minutes. Then resume rubbing. It is important to rub, rub, rub all over. Let it remain another five minutes, then remove with a damp washcloth. Finish with a lukewarm water splash, a cold water splash, and finally a raw fruit juice splash.

*Enzyme Benefit:* The enzymes in the avocado combine with this plant's oils to blend with the grains of the cornmeal. The action of your fingertip rubbing activates the enzymes to create a natural "scrubbing" over the skin surface, and deep down within your skin pores. Here the plant enzymes work with the gentle abrasive cornmeal grains to help lift up and slough off dead tissues. It is Hollywood's all-natural secret face scrub that can help give you the "forever young" look of a film star, thanks to plant enzymes.

### *How Enzyme Juices Help Tighten the Skin of the Neck and Jawline*

Adelle F. had to wear high collars because she was ashamed of her sagging neck and jawline. She had tried one cream after another, and while her skin felt better, the effects were temporary. Adelle F. realized that most commercial preparations offer a brief softening while she wanted something more effective and longer-lasting. A natural food enthusiast friend told her that just as enzyme juices help to build health from within, they could also be effective in helping to build health from the outside. The friend had a remarkably clear skin, firm throat lines, a smooth and young looking face—even though she was older than Adelle F. The friend gave Adelle F. her special all-natural enzyme program for external rejuvenation. Here is the program as Adelle F. followed it:

**Enzyme Rejuvenation Program for Neck and Jawline.** Mash half an avocado and stir into it one tablespoon of fresh cold-pressed wheat germ oil. Rub this cream into the neck and jawline areas with firm up and down strokes. Keep rubbing. If desired, rub around the nose, mouth and the lines of the face. When most of the cream has been absorbed, it indicates that enzymes in the avocado and the cold-pressed wheat germ oil have been absorbed. Now remove the excess. If possible, let it remain overnight.

*Secret Working Power of Rejuvenating Enzymes:* These miracle youth workers worked overnight. They supplied to the epidermis layer of the skin a synthesis of stimulation that helped furnish concentrated moisture to help plump up starved cells beneath. The enzymes also awakened sluggish enzymes beneath the skin surface to create a rehydration process that helped to fill up the thirsty tissues that had previously become dried and scaly. While Adelle F. slept, the enzymes performed their secret powers of invigorating sluggish cells and tissues beneath the top layer of the skin surface.

**One Week to an Elegant Neck and Jawline.** It took just one week of this nightly enzyme treatment before Adelle F. noticed how her sagging throat was now firm and tight. The enzymes had helped firm up the flabby jawline, tighten up the sagging muscles, and smooth up the crepey neck. Now Adelle F. looks and feels much younger, thanks to this all-natural "rejuvenate-while-you-sleep" enzyme treatment.

### *Parsley-Peppermint-Lettuce Facial Lift*

A youthful skin needs *living* tissue. Enzymes are living substances that help nourish and boost body cells and tissues and give them new life. The following program contains fresh and living substances that are chock full of enzymes that go to work speedily to help alert and awaken your sluggish skin cells and tissues. Reportedly, the nightly use of this Parsley-Peppermint-Lettuce Facial will give all the appearances of a "natural face lift."

Mix two pinches of fresh parsley, two pinches of peppermint leaves, two pinches of dried rose hips, and two pinches of dried sage with a teaspoon of hot water. Then add a teaspoon of fresh lemon juice. Meanwhile, place three cucumber slices and three potato slices in a blender, add the herbal mixture and whir at the low speed. When the mixture is creamy smooth, pour into a cup. Now immerse one side of the lettuce leaves in the mixture and apply directly to your face and throat. Keep using the lettuce leaves for applying the mixture to your skin. Let remain up to an hour. Rinse off and then splash with lemon juice. (*Tip:* For more effectiveness, steam the lettuce leaves with hot towels while they are on your face and throat.)

*Enzyme Facial Lift Secret:* This natural face lift program reportedly works because its substances are nonspecific biogenic stimulators. The secret is that they have a general ability to stimulate the metabolism of your skin's living cells and to act therapeutically upon them by raising their functional response. Once this is done, the stimulators then alert the enzymatic processes located right beneath your skin surface. This helps raise the tonus of your epidermis and subcuteaneous tissue to reinforce their health. Now the skin becomes firm, clear and more youthful in appearance because the enzymes from the aforementioned ingredients have created this biogenic-cosmeto-therapeutic action within the skin cells and tissues. Many have found this all-natural program to create a look and feel of a "face lift" in the privacy of their homes.

### *How to Take a "Plant Sauna" Face Treatment*

Plant enzymes can work with double-action speed if they can get into the skin pores with ease. You may have heard of the Finnish sauna bath in which dry heat is used to open the skin pores and promote natural cleansing. You can take a "Plant Sauna" face treatment right at home, *without* a costly sauna appliance, by using this method:

Fill your wash basin with steaming hot water. Cover your head with a towel "curtain" so the steam does not escape. Hold your face about 12 inches above the water to prevent the heat from being too intense. Experiment until you find the most comfortable position for you. Let the sauna steam open your pores and cleanse away clogged impurities and age-causing impediments. About 30 minutes should be sufficient.

Next, just pat your face dry with a Turkish towel and apply the mashed pulp of any fruit or vegetable. Let this pulp remain for about one hour. The benefit here is that once the skin pores have been opened by the sauna heat, the enzymes from the plants are able to enter very speedily. Once the enzymes have penetrated, they immediately replenish nutrients and alert and nourish the subcutaneous tissues and cells. The enzymes assist in the replacement, retention and increase of the skin's natural oils and moistures.

A special benefit is that the plant enzymes are remarkably compatible with the natural fluids of the skin; enzymes send plant juices to merge with existing skin juices to help boost the dwindling cellular levels so that the skin is now encouraged to regain its youthful bloom.

Plant enzymes also help maintain the delicate supplies of moisture needed to have a dew-fresh appearance of a youthful skin. Given a needed supply of this rich moist oil from the plant, transported by enzymes, the skin is able to resist the inclination to dehydrate or look prematurely or unnecessarily aged. Plant enzymes send moisture within the skin follicles to aid in maintaining resilience and a young-feeling texture. The end result is a peak-of-perfection softness. Your skin now enjoys rightful softness and smooth "kiss of Nature" youthfulness. Plant juices are transported into the skin pores by the propelling of the enzymes. To use the Finnish sauna for a few moments is to open the gateways of your pores to a new horizon of youthful skin health.

### *A Grain Scrub Creates Glowing Skin*

Barbara P. used fruits and vegetables to help "feed" her skin with natural masques. But the results were still short of her expectations. She then improved her program by *first* scrubbing away debris and clogged-up wastes. Barbara P. learned that plant enzymes are able to work more efficiently if the skin pores have first been cleansed and opened to admit these miracle youth workers.

Barbara P. used a simple but highly effective grain scrub to cleanse her clogged pores. She mixed almond meal with enough water to make a paste. She applied this "cream" to her entire face and let it remain up to 30 minutes. Ingredients in the grain-like almond meal worked to "suck up" clogged impurities. Barbara P. then washed it off with warm and cold water, and used a lemon juice rinse to restore the natural acid mantle.

She was delighted to note that this Grain Scrub helped relax her tired facial muscles and ease the crease-tension lines on her face and throat. She saw blackheads and other impurities actually come off. Her "open" pores were now receptive to the enzymes in the creamed pulp of any fruit or vegetable used as a masque. This simple program of a grain scrub to open the pores and a plant masque to send enzymes into the skin to help replenish decaying cells did help give Barbara P. a firm, tight, clear and youthful skin. It was the closest thing to "plastic surgery," she would boast, done right at home, using plant enzymes!

### Potato Enzymes Help Soothe Puffy Eyes

The potato is a prime source of Vitamin C which is used by enzymes for helping to nourish the fragile capillaries of the body with *collagen,* a healing substance that binds and repairs broken cells and tissues. Potato enzymes are also valuable for helping to soothe puffy eyes that often occur after much reading or sleeplessness.

*Potato Enzyme Healer:* Grate three thick slices of a peeled raw potato. Wrap in two separate pieces of cheesecloth. Now warm any seed or vegetable oil over a low flame. Spread this warm oil around your eyes and on your eyelids. Lie down. Now cover each eye with a cheesecloth potato sack. While you rest, the enzymes of the potato will seep out, enter the favorable oil-anointed environment of the skin around your eyes, penetrate beneath and slowly help replenish the damaged capillaries with much-needed Vitamin C from the potato. Enjoy this eye-healing home treatment up to 30 minutes. Then remove the potato sacks and discard. Rinse your eyelids and the skin beneath your eyes with tepid and then refreshingly cool water. Gently dry yourself.

### Papaya Enzyme Pep-Up

As an interior decorator, Irene W. is always on the go. Frequently she begins work early in the morning in her clients' homes, and spends much of the afternoon in decorating showrooms making selections. She then returns to her clients' homes in the evening to discuss her ideas and plans. Not only must she display boundless energy, but also look youthful and alert. It is part of her image in the decorating profession.

Irene W. follows the natural health program of eating lots of raw fruits and vegetables to give her body the enzyme working materials for better health. She does even better by using plant enzymes on her face and neck so that they can help promote the same look and feel of youth as they have done from within. She manages to find just 30 minutes a day for a simple "pep-up" program. Here it is:

*30-Minute Enzyme Pep-Up:* Combine a cup of mashed fresh papaya with four teaspoons of unflavored yogurt. Blend very thoroughly. Now massage gently onto the face, neck, and soft hollow of the throat. Let it seep and soak through the pores of the skin for about 20 to 25 minutes. Now remove with splashes of warm water. Follow with some cool water. Then a sprinkle of lemon or citrus juice to restore the needed protective acid mantle.

*Youth Benefits:* The papaya in this 30-Minute Enzyme Pep-Up program contains papain, mylase, prolase which have powerful protein, starch-dissolving properties. These three enzymes are propelled by the acid of the yogurt to seep through the pores, to help "lift" away the accumulations of dead outer skin particles. The benefit is that these enzymes help "aerate" the skin, promoting better "ventilation" and "oxygenation." This speeds up a sluggish circulation and helps promote a better assimilation of nutrients. The end result is a feeling of youthfulness from within and a look of youthfulness from the outside.

Irene W. is an effervescent interior decorator who is a living example of youthful color and attractiveness, thanks to the rejuvenating powers of enzymes as well as her daily 30-Minute Enzyme Pep-Up program.

Raw fruits, vegetables and grains have Nature-created enzymes to promote living health to the body. To enjoy a feeling of youth, eat enzyme-rich natural raw foods. To enjoy a look of glowing youth, let your skin "eat" enzymes to nourish your cells and tissues so that you help create the bloom of health from head to toe.

### *Main Points:*

1. Living enzymes in living plant and grain foods help "feed" your skin to promote sparkling youthfulness.

2. Ordinary lemons are rich in enzymes that help replenish "tired skin" from head to toe. It helped Martha B. enjoy a new "bloom of youth" so that she could be mistaken for her husband's younger sister.

3. Citrus fruit enzymes create a proper acid balance of

the skin mantle to give it a clear glow. Fruit enzymes made Sophie D. look "too young."

4. A Berry-Rose Enzyme Skin Pickup has a natural humectant action to moisturize the parched cells and tissues beneath the surface.

5. A special Hollywood Secret Face Scrub is now available to you right in your own home with just three items available at any store. Creates a "million-dollar" youthful skin.

6. A simple "Plant Sauna" Face Treatment helps roll back the years in just moments.

7. Pep up your face with a 30-minute enzyme treatment using just three ingredients found everywhere at low cost. Irene W. bounces back with youthful vitality with this simple program.

# 4

## How Europeans Use Enzyme Oils to Help Melt Arthritis Stiffness

The mysteries surrounding arthritis are being penetrated by European healers with the use of rejuvenating enzymes found in the living juices of seeds and plants in the form of cold-pressed oils. It has been reported that a corrective eating program that emphasizes the intake of fresh raw foods and the regular use of cold-pressed oils offers the body a good supply of restorative enzymes that have been seen to bring about a "melting" of arthritis-like stiffness of swollen and inflamed joints.

### *The Natural Healing Benefits of Enzyme Oils*

It is reported that living enzymes in raw foods as well as in cold-pressed oils will help create these basic natural healing benefits for those troubled with inflamed swollen joint stiffness:

1. Enzyme oils give the body the defensive and healing forces that help to correct metabolic disorder and systemic disturbances traced to arthritis. Enzymes work to create order in the arthritic's body, to restore vitality to his vital organs, to improve the digestive powers, to help balance endocrine glandular secretions and boost hormonal supply. All of these functions are seen to be disordered in the arthritic. Enzyme oils help restore internal balance so that healing forces may then enable the body to recover from arthritic-like disorders.

2. Enzyme oils alert the body's own healing processes to stabilize metabolism. The enzyme oils enter into all functions

to rebuild the individual, to enable his own healing powers promote a form of natural restoration of better health. In particular, enzyme oils help promote healing of the body tissues, especially collagen (the healing cement-like substance of cells and tissues in the bone marrow) so that the so-called stiffness of joints can then be "melted" through a natural recovery.

Enzyme oils alert a sluggish metabolic system so that it can then help create its own form of easing of discomfort and a slow and gradual form of healing. Enzymes in plant and seed oils are *living organisms* that help bring new life to internal decay.

### *Cold-Pressed Oils Are More Healing*

Of all the methods of extracting oils from nuts, grains, beans, seeds, and vegetables, the cold-pressed method is seen to be the most beneficial for its supply of healing enzymes. A hydraulic press is used to extract the oils from the foods. Because no heat is used, the enzymes are preserved, along with Vitamins A, D and E. These vitamins work in harmony with the enzymes to help create better health for the body. Cold-pressed oils are as close to Nature as possible and therefore have the most color, flavor and enzyme-nutrition. Europeans prefer the use of cold-pressed oils for healing purposes. They have found that arthritic-degenerative ailments are traced to disordered or impaired metabolism caused by nutritional deficiency. To use natural, wholesome raw foods as well as cold-pressed oils which are prime sources of enzymes and nutrients is to give the body the raw working materials from which health rebuilding may be created from within. Enzyme oils form the foundation for this healing.

### *How Enzyme Oils Helped Soothe "Restless Legs"*

In a reported case, several persons complained of nocturnal leg cramps or arthritic-like stiffness. The persons were unable to sleep because of the pain. Many resigned themselves to their fate. But others sought help from natural enzyme oils and were able to "melt" arthritic-like stiffness of such pains.

**Enzyme Oils Relieve Leg Pains.** Charlotte E., at age 78, had more than just arthritic-like leg cramps. She complained of "restless legs" or "jumpy legs" which had bothered her for twenty years. Charlotte E. could hardly get enough sleep. She was placed on a simple yet effective program of raw enzyme oils. In a short time she was able to enjoy relief and freedom from this type of arthritic-like distress. Enzymes in the oil had helped restore better metabolism and promote soothing relief from within her system.

**Rheumatism is Eased with Enzyme Oils.** In another reported situation, some 300 persons had problems of *fibrositis* (inflammation of the white fibrous connective tissues that form muscle sheaths and merge into muscle attachments) and were put on a special program that emphasized enzyme oils. It is reported that a majority of these people were able to become healed of *fibrositis* because of the enzyme activity of cold-pressed oils.

### *How Enzymes in Wheat Germ Oil Helped Heal Arthritic Distress*

Cold-pressed wheat germ oil, available at almost all health food stores and many large food outlets, is a prime source of enzymes that help correct metabolic disorders and promote soothing relief of arthritic-like symptoms. Four reported cases illustrate[1] the healing power of enzymes in wheat germ oil:

1. *Housewife.* Her hands were swollen and stiff. The use of cold-pressed wheat germ oil helped return her hands to their normal condition.

2. *Physician.* Arthritis had stricken this physician in his knees and back. The swelling was so painful, he could not walk. He was put on a special enzyme program that called for the regular use of cold-pressed oils and was soon able to resume his practice. His knees and back felt youthful and supple again.

3. *Elevator Operator.* A serious injury gave him pain and stiffness in his chest and legs. He developed bronchial disorders. He was put on a simple enzyme program calling for wheat germ oil to be taken daily. After several months he

1. Morris Ant, M.D., "Wheat Germ Oil, and Arthritis," *Industrial Medicine* (6:46). pp. 48-50.

returned to work free of all arthritic-like symptoms, except for a slight occasional limp. His bronchial disorder and asthma cleared up.

4. *Store Clerk.* When she fell against the sharp corner of her desk, she suffered a severe pain in her ribs. It was diagnosed as a form of arthritis. It grew so painful that she was unable to sleep. After being put on a special enzyme program, she took wheat germ oil regularly. Within a few months all symptoms vanished and she felt flexible all over.

Cold-pressed wheat germ oil is just one of many available oils that are prime sources of vitamins as well as enzymes. Europeans have long recognized that if the body is to be able to resist and recover from disorders of arthritis, it must have youthful metabolism. Many European countries have long winters and are deprived of sunshine as well as the availability of fresh raw enzyme-rich fruits and vegetables. Because of this, they have come to depend upon cold-pressed plant and seed oils for a steady supply of rich nutrients that will help keep them healthy and fit during long winter confinement.

### *How Enzyme Oils Help Create Youthful Metabolism and Freedom From Arthritis*

Enzymes in plant and seed oils influence the utilization of Vitamin D, calcium and the parathyroid hormone. (Parathyroids are several small glands located behind or embedded in the thyroid in the front of the neck.) When you have sufficient enzymes from plant and seed oils, they help create these youth-building metabolic actions that offer hope for freedom from arthritis:

1. Enzyme oils alert the parathyroid glands to issue forth their hormones to send calcium to the bones, strenthen the skeletal structure, build fibers that help guard against spontaneous fractures, and help strengthen the muscles to resist problems of weakness, swelling, tenderness or pain.

2. Enzyme oils prompt the parathyroid hormone to promote the passage of calcium across the intestinal mem-

brane. Enzyme oils use calcium for the building of bones and teeth, maintenance of blood coagulation and cardiac rhythmicity, regulation of membrane permeability, and maintenance of neuromuscular excitability and muscular contraction.

3. Enzyme oils use Vitamin D to create normal maturation and growth of epiphyseal cartilage (from which bone grows), for osteoblastic activity (the actual growth of bone) and the arthritis-healing process of "ripening" of osteoid seams (the laying down of bone in an architectural pattern which corresponds to stress lines).

4. Enzyme oils prompt calcium to become absorbed by the body. Once absorbed, calcium then joins the extracellular "pool" in the form of calcium phosphate. The calcium in this pool is then either put into the intracellular "bone bank," or excreted via the kidney or intestines. Enzymes work to keep the balance in the pool steady so that the calcium ions perform all their necessary functions. If the calcium in the pool is low, the bone bank contributes to the extracellular pool.

5. Enzyme oils help maintain a smooth-functioning parathyroid gland, a key to healing of arthritis. If there is a deficiency of enzymes, the parathyroid glands may become either overactive or underactive. If *overactive,* phosphate and calcium are lost via the kidney, causing a relative undersaturation of the calcium in the extracellular pool. The pool must then drain more calcium from its sources of supply such as the bone, and arthritic disorders may follow. If *underactive,* calcium cannot be absorbed and phosphate cannot be excreted from the kidney. This leads to a *rise* in serum phosphate and a *fall* in serum calcium. The pool is depleted of calcium. As a result, all the functions of calcium are interfered with and the skeletal structure which depends upon calcium sent to it via enzymes begins to suffer. Arthritic symptoms then develop. With the presence of enzyme oils, this process could be stabilized and balanced. Enzymes help maintain a natural removal-replacement rhythm of minerals to create strong bones and bone marrow to help resist any possibility of arthritic unrest.

### Why Polyunsaturated Oils Are More Beneficial Than Saturated Fats

*Polyunsaturated oils* (particularly the cold-pressed oils) are prime sources of living enzymes as extracted from the plants and seeds from which they are made. The cold-pressing process does not destroy the enzymes which are heat sensitive.

*Saturated fats* are in the form of butter, margarine and animal fats. Heat processing and preserving are destructive to the enzymes so that the "hard" fats create cholesterol deposits that act as barriers to the free movement of body enzymes as well as other essential nutrients. Saturated fats leave plaques that block the natural circulatory processes and inhibit the bone-strengthening powers of vitamins, minerals and enzymes.

European healers have long known that the use of as many polyunsaturated fats (enzyme-rich plant and seed oils) as possible, can help create better metabolism and resistance-recovery to arthritic distress.

### The All-Natural English Food Program That Helped Heal Arthritis

In a reported study, a British physician selected 12 patients who were diagnosed as having chronic rheumatism. Symptoms varied from muscular arthritis to osteoarthritis and rheumatoid arthritis. Reportedly,[2] the patients were given enzyme-rich *raw foods,* with emphasis on raw plant oils. Here is a sample two-week raw food program as prescribed by the doctor:

*Breakfast:* Applesauce made from grated apple, soaked raw oatmeal, grated nuts, cream, fresh orange, tea with milk and cream.

*Mid-morning:* Mashed tomato with lemon juice.

*Dinner:* Raw vegetable salad of lettuce, cabbage, tomato, root vegetables with lots of fresh cold-pressed oil. A mixed fruit salad.

---

2. Dorothy C. Hare, M.D., "Raw Food for Arthritics," *Proceedings Of The Royal Society of Medicine,* Vol. XXX, pp. 59-63.

*Mid-afternoon:* Sun-dried fruits, nuts and tea with milk and cream.

*Bedtime:* Lemon and orange juice with hot water.

*After two weeks:* Vegetable soup, one egg, two ounces of meat, two ounces of bread, cheese and milk.

*Note:* The emphasis here is on the use of raw enzyme-rich foods. The cold-pressed oil that was a "must" for dinner took up the vitamins and minerals from the raw foods and used them for the building and rebuilding of the body.

*Results:* Eight of the arthritics felt better in one to four weeks on this enzyme-rich all-natural food program. There were some setbacks, and a few of the arthritics showed no response.

**Bedridden Arthritic Walks Out of Hospital.** At age 46, Elizabeth H. was a bedridden arthritic. She had pain and stiffness in her shoulders, arms, hands, knees and legs. She was diagnosed as having fluid in both knee joints, and swelling and pain in other joints. For ten weeks Elizabeth H. was a bedridden arthritic.

She was placed on this enzyme-rich program with extra emphasis upon cold-pressed oils. Just three weeks of this simple yet effective enzyme-rich program and Elizabeth H. could walk out of the hospital. She continued on the program at home and reportedly was free from arthritic pain and able to do her housework.

*Secret of Success:* Cold-pressed plant oils are prime sources of enzymes which help stir up sluggish circulation and send these metabolic agents sweeping throughout the body to send nutrients to "starved" joints and bone pockets to create healthful healing. With a raw food program, there is a double action of enzymes introduced into the body to help increase the effective working vigor of enzyme oils.

### *The German "Enzyme Butter" and Its Healing Benefits*

At the Otto Buchinger Sanatorium in Bad Pyrmont, Germany, arthritic patients are immediately placed on a raw food diet, with the stipulation that only plant or seed oils are to be used for fats. Many arthritics are "addicted" to hard fats and crave the taste of rich, creamy butter. To help ease the transition, the Buchinger

Sanatorium has created a healthful "enzyme butter" that has all of the taste of the saturated butter but is also brimming with enzymes. (Because butter is prepared by heating processes and is also a saturated or "hard" fat, it is void of life-giving enzymes. The Buchinger doctors have been able to put enzyme "life" into "lifeless" butter with this make-it-yourself recipe.)

**The German "Enzyme Butter" You Can Make at Home.** When fats are hydrogenated, the hydrogen is added to the delicate essential fatty acids, thus destroying their health-building values. Added hydrogen also is destructive to enzymes that may have originally been present in the ingredients that went to make up the butter.

At the Otto Buchinger Sanatorium, healers prepare a popular German "Enzyme Butter" which can be easily spread, is less likely to turn rancid, and is a good source of enzymes and essential fatty acids.

Let one pound of butter soften at room temperature. Now mix this one pound of softened butter with one cup of cold-pressed oil. You may whip together in an electric blender, or do it by hand with a wire whisk.

When thoroughly blended, pour the mixture into leftover margarine tubs or glass bowls, cover and refrigerate. The "butter" will now harden and be white in color, but will keep its taste, fragrance and spreadability when cold.

At the Otto Buchinger Sanatorium they make large batches of this "Enzyme Butter" by doubling or tripling the amounts of butter and oil. Keep the "Enzyme Butter" refrigerated since cold will inactivate but not destroy enzymes. It is the natural way to get all of the taste of rich creamy butter, but with less of the saturated (dead enzyme) fat and with much more of the essential unsaturated (living enzyme) fatty acids.

*Suggestion:* You may add diced vegetables to this European recipe before freezing to further increase its enzyme content. The vegetables should, of course, be fresh and raw.

### *How Swiss Healers Use Enzymes to Help Relieve Arthritis*

At the Bircher-Benner Clinic in Switzerland, a program of natural raw foods, together with plant and seed oils, forms the basis for helping to correct metabolic error and promote relief and healing of arthritis.

### *The Ten-Point Swiss Program for Enzyme Healing of Arthritis*

Adequate nutrition is important for helping to build the health of the arthritic. The Swiss healers say it takes from a week to a month before the body begins to experience some recovery and slowly correct biological defects and begin the process of healing. The ten-point Swiss program is simple, easy, natural and reportedly effective.

1. Every day, eat a lot of fresh raw fruits and fresh raw vegetables. The Swiss healers report success when more than 50 percent of the daily food intake consists of fresh raw fruits and fresh raw vegetables.

2. Fresh raw seeds (especially sunflower, sesame and pumpkin seeds), nuts, wheat germ and brewer's yeast should also be taken throughout the day. The benefit here is that enzymes in these seeds are potent carriers of Vitamins A and D that are needed to build and rebuild the bone structure and surrounding tissues that have become degenerated due to arthritis.

3. All foods should be *without* sugar in any form. The flour products should be made from whole, unbleached and nonprocessed flour.

4. As many vegetables as possible should be eaten raw. If vegetables must be cooked, use the minimum amount of cooking time to avoid loss of vitamins, minerals and valuable enzymes.

5. Whole grain breads and whole grain cereals are sources of good nutrition. They also have essential fatty acids which are "rivers" upon which enzymes "swim" toward their healing objective.

6. Avoid any fried foods, preserved foods, chemical preservatives, sprays, artificial sweeteners, coffee, tea, cocoa, chocolate, carbonated beverages. The processing has

destroyed the enzyme content and these are regarded as "lifeless" foods by the Swiss healers. Substitute with natural, nonprocessed foods. Drink herb teas.

7. Begin each meal with an enzyme-rich raw salad, fresh vegetable juice or raw fruit. The Swiss healers often suggest using lots of cold-pressed plant or seed oil upon the rich raw vegetable salad as a means of boosting the enzyme intake into the system. Enzymes obtained in this manner at the *start* of a meal help favor better digestion and assimilation.

8. Enzymes work with proteins (since enzymes are protein-like substances) to promote internal healing. The Swiss healers suggest eating protein as found in fresh, green leaves, whole grains, and seeds. One protein complements another and strengthens the enzyme power in the system.

9. A higher enzyme value is found in fermented foods and beverages such as certified raw milk, buttermilk, yogurt, some cottage cheese and natural cheeses. The Swiss healers shun the use of homogenized-pasteurized milk which has very low (if any) enzyme content.

10. A vegetable juice becomes a powerful Enzyme Tonic when two tablespoons of nonprocessed, cold-pressed wheat germ oil have been added. The benefit here is that the enzymes in the raw vegetable juice are electrified by the enzymes in the wheat germ oil and then speedily absorb Vitamins A and D in the system to be used for the healing of the bone structures adversely affected by arthritis.

The famed Bircher-Benner Clinic of Switzerland reports the healing of many cases of arthritis on a program of good nutrition that emphasizes the intake of cold-pressed oils.

### *The European Waerland System for Arthritis Healing*

The Swedish nutritionist and health pioneer Are Waerland created a system of healing that was based upon this single premise:

*Arthritis is not a disease but a symptom of errors in living. Correct the errors and the symptoms will vanish along with the cause. The road to arthritis healing is in the rebuilding of the body with natural health habits.*

Dr. Waerland reportedly healed tens of thousands of arthritics with a natural program that emphasized using cold-pressed oils that acted as carriers of enzymes to promote healing from within. He helped establish healing spas which followed his programs with remarkable success. He describes[3,4] his programs with reported successes as follows:

**Morning Enzyme Tonic.** (This should be prepared the night before.) Soak a tablespoon whole flaxseed with a tablespoon wheat bran in one cup of raw vegetable juice. Let it soak overnight. In the morning, sprinkle with some raw wheat germ. Add a teaspoon of honey to taste. Drink one cup on arising.

*Enzyme Benefit:* The flaxseed and wheat bran are natural supplies of enzymes. These enzymes combine with alkaline substances in the grains and help to neutralize excess body acid. The enzymes work with the oils to help wash out abrasive or corrosive metabolic wastes that are harsh and volatile. The enzymes "ride" upon the oils to transport needed vitamins and minerals to the bone structure to help bring about inner healing.

**Natural Breakfast.** The Are Waerland system calls for a breakfast consumed shortly before noon. The benefit here is that by noontime, the bloodstream has been cleansed of the waste products of metabolism via a natural elimination. This will enable food enzymes to work without interference in helping to rebuild health.

An enzyme-rich breakfast consists of fermented milk (such as yogurt or buttermilk) and fresh raw fruits.

*Suggestion:* Stubborn cases of arthritis seem to respond to this enzyme-charger: stir finely chopped raw onions into a cup of yogurt. Sprinkle wheat germ or milk sugar (lactose) on top, and eat slowly with a spoon. The enzymes of the fermented milk work to extract the enzymes from the raw onion and help give the body a super-powerful supply of health-building factors. Enzymes from this simple mixture appear to boost sluggish metabolism and awaken weak digestive functions to create better assimilation of vitamins, minerals, essential fatty acids and proteins.

---

3. *Rebuilding Health,* Sweden: Ny Nords Forlag.
4. *Health Is Your Birthright,* Bern, Switzerland: Humata Publishers.

**Snack Time.** Fresh fruits are suggested since they are potent sources of natural oils that are needed by the body.

**Luncheon.** Mix together equal portions of whole grain and nonprocessed sunflower seeds, wheat germ, oats, barley, millet, and wheat bran. Stir into a cup of fermented milk. Sprinkle with sun-dried fruits of any sort and raw wheat germ. Eat with a spoon.

*Enzyme Benefit:* The oils from the grains, seeds and fruit are introduced into your body in a natural and wholesome form. They have not been processed or tampered with. Just as the enzymes in the grains, seeds and fruits nourish these plants, so they reportedly will nourish your body from within. Oils are extracted from the ingested seeds by the process of metabolism, and they speedily act as transporters of nutrients to all reaches of the body. The *secret* here is that they help to correct the metabolic defect that is symptomatic of arthritis.

**Dinner.** The Waerland System calls for beginning with a large raw green leafy vegetable salad with lots of cold-processed oils. Add any combinations of available seasonal raw vegetables you wish.

The Waerlands report that enzymes are especially beneficial to the arthritic if introduced in the form of three vegetables: onions, carrots, and beets. They must be raw, of course.

In addition to this vegetable and oil salad, eat whole grain bread, together with cottage cheese. They advise fermented milk with this raw vegetable meal.

*Enzyme Benefit:* The acidophilus bacteria and enzymes from the fermented milk work upon the tough fibers of the raw vegetables to extract their valuable healing oils. The enzymes then send these oils propelling throughout the system to promote better metabolism and assimilation of nutrients. In particular, the oils help to nourish the bone of the arthritic. The enzymes use the oils for gathering up of nutrients to create oxygenated compounds of calcium and phosphorus: this is the mineral mass of the bones. The enzymes from the fermented milk use the oil from the raw vegetables to create a healing removal-and-replacement balance of calcium and phosphorus compounds. This is the physiological process that promotes strength and flexibility of the joints. It is this balance that is needed to ease arthritis. The Waerland System, with its emphasis upon raw foods for enzymes and cold-pressed oils for transport of nutrients, is aimed at restoring this needed physiological process.

**Do's and Don'ts.** At the Waerland healing sanatoriums, all foods that can be digested without being cooked are eaten raw. If cooking is necessary, a minimum of steaming is done to make the foods palatable. This helps preserve the supply of enzymes.

No volatile spices or condiments are permitted. Salt, pepper, mustard, catsup, and artificial seasonings are taboo. These are synthetic chemicals which destroy or inhibit enzyme content in the system.

Beverages should be fresh raw fruit or vegetable juices. No coffee or tea are permitted, except herbal tea.

Alcohol and tobacco are forbidden.

White sugar and white flour are also taboo, along with anything made with these such as cakes, candies, pies, desserts, ice cream, chocolate, or confections.

All foods should be as natural and nonprocessed as possible. The Waerlands prohibit the use of any refined, canned, bottled, processed or adulterated foods.

**How the Waerland System Healed Monica W. of Lifelong Arthritis.** Before she was three years old, Monica W. was stricken with arthritis. She was sent from one hospital to the next, but received little in the way of healing. Her condition became progressively worse. When she went to school, her painful joints were so twisted that she had to be carried home. Her parents, hearing of the Waerland System, sent her to the clinic for treatment. No drugs were used. Here is the program that healed Monica W. of lifelong arthritis:

1. A one- or two-day fast during which the only nourishment is some water to which the juice of a half lemon is added. This helps cleanse the system of accumulated debris and impurities.

2. On the first day of food intake, the menu calls for raw fruits and vegetables with lots of seed or plant oil for dressing. Fermented milk is taken with these raw foods.

3. The aforedescribed food program is followed.

4. Daily, Monica W. was given a self-massage with a rough cloth or a dry soft brush to stimulate cells and tissues

to cast off metabolic toxins and body wastes from the system. In many arthritics, impaired circulation means that the skin (a body organ) has become inactive. Your body skin's pores become clogged with millions of decaying cells, wastes and debris. In Monica's case, years of sluggish arthritis had choked up her pores so that her insides were further clogged with wastes that could not be released. Her daily brushing (with either a natural bristle brush, a rough bath glove, a plant fibre brush or a coarse sponge) helped boost sleepy circulation. The brushing from head to toe helped slough off decaying skin cells and open the pores. Once her pores could breathe, then accumulated toxins could be eliminated. The brushing also helped stimulate the glands that produce natural oils. Slowly, she underwent simple but healing autolysis–self-loosening and discarding of accumulated corrosive wastes.

5. A daily contrasting hot and cold shower was given. The comfortable temperature extremes create a gentle nudge to the metabolic processes. Now the enzymes are whipped up into healthful activity so they can send nutrients to the ailing and starved organs to help promote the building and rebuilding caused by the ravages of arthritis.

6. Emphasis is placed upon the complete elimination of artificial foods. All foods should be as natural and wholesome as possible.

7. Nightly rest is essential.

8. To help send vitamin-carrying enzymes throughout the system, Monica W. was allowed to munch freely on raw nuts, raw seeds and sprouted grains. This helped Monica's body become cleansed through the scrubbing action of the enzymes. Furthermore, the oils in the raw nuts and seeds and grains helped lubricate the joints. An added benefit was that these oils acted as carriers of vitamins and minerals and some amino acids that work together to promote effective healing of arthritis-injured internal organs.

*Healing Miracle:* Monica W. underwent the all-natural program as described above for six weeks. The Waerland healers examined her, pronounced her "completely healed of arthritis" and released her from the sanatorium. Monica W. was later examined and found to be completely healthy. Arthritis never returned. She was now normal, healthy and vigorous, thanks to the all-natural enzyme healing powers of the Waerland System.

### *The European Discovery About the Arthritis-Healing "EFA" Factor*

European healers have successfully isolated cold-pressed oils from plant and seed foods and have dubbed them as arthritis-healing "EFA" factors. They are *essential fatty acids,* a type of linkage in the fatty acid where two adjacent carbon atoms are linked by two bonds. This linkage is a site of unsaturation. Essential fatty acids contain more than two such sites of unsaturation per fatty acid molecule.

The "EFA" factors found in cold-pressed oils from plant and seed foods are beneficial toward the metabolic process for the arthritis in that they have the ability to take up vitamins A, D and E as well as valuable minerals such as calcium and phosphorus, and transport them to needed "starved" or "ailing" organs and joints that have been struck by the ravages of arthritis.

In contrast, the "hard" or "saturated" fat found in animal products works less efficiently, creates deposits of cholesterol that impede and obstruct the free-flowing ability of enzymes. In many arthritics, there is an elevated blood cholesterol level that reportedly becomes reduced as the "EFA" factor is introduced.

*Where to Get the Arthritis-Healing "EFA" Factor:* Any cold-pressed oil from a nonanimal source (except olive oil) is said to be rich in the essential fatty acids that act as carriers of healing enzymes and nutrients. Europeans use corn oil, peanut oil, safflower oil, sesame oil, soybean oil, sunflower oil, almond oil, avocado oil, linseed oil, rice bran oil, walnut oil, and wheat germ oil, among others. Many of these oils are available at almost all large supermarkets as well as health food stores throughout the country. Use these oils in place of "hard" fats and give your body the valuable arthritis-healing "EFA" factor as do the Europeans who may be considered far ahead of us in the discovery that enzymes in oils hold the key to youthful health.

*In Review:*

1. Europeans have reported that enzyme oils help promote natural healing of arthritis through correction of metabolic disorders.

2. Enzyme oils helped soothe "restless legs" and also reportedly healed arthritis in other victims.

3. Enzyme oils have miracle restorative powers in correcting internal upsets that are related to arthritic flare-ups.

4. An all-natural English food program reportedly healed arthritis and rheumatism in "hopeless" cases.

5. The German "enzyme butter" is a prime source of nutrients needed for internal rebuilding.

6. The Swiss healers use a ten-point all-natural program for enzyme healing of arthritis that can be done right in your own home.

7. The European Waerland System is simple yet effective in promoting correction of metabolic error and healing of arthritic symptoms.

8. Monica W. experienced lifelong cure of arthritis with this all-natural program.

9. The "EFA" Factor in cold-pressed oils acts as a transporter of nutrients to "starved" organs and joints in need of nourishment for rebuilding and healing of arthritis.

# 5

# *Oriental Secrets of Youth-Extending Enzymes*

The Orientals are known for living well beyond the century mark in seemingly good mental and physical health. For thousands of years, the Orientals have enjoyed an appearance of perpetual youth, in the absence of conventional medicine as we know it today. From the legendary eternal-living Tibetans to the modern youthful Orientals, the people of the Asiatic region appear to have tapped the secret of rejuvenation from a mystical source. Modern scientists have researched the oriental secrets of perpetual youth from scrolls handed down for thousands of years right up to the present time. They have learned that the Oriental believes in using all-natural methods for *stimulating* and *reviving* his own built-in resources of healing and rejuvenation. It is more than just a necessity in the absence of conventional medicine. It is a tradition that reaches back in the mysterious heritage of oriental secrets of youth-extension.

## *The Yellow Emperor's Five Secrets of Eternal Youth*

The first written work known to contain the secrets of eternal youth is *The Yellow Emperor's Book of Internal Medicine (Huang-ti nei Ching),* prepared by Emperor Shen-Nung (the inventor of the plough). In this classic book, the Emperor described the secret based upon the principles of *Yin* and *Yang.* Written over 4000 years ago, the book explains that self-rejuvenation and self-renewal could be possible when the opposing elements of Yin and Yang could be brought into harmonious balance. These two principles may be likened to the modern awareness of enzymes. Namely, when the enzymes in the digestive and assimilative systems are properly balanced, there is hope for better health.

Emperor Shen-Nung explained that certain digestive elements must be in harmony if they are to benefit from foods that are eaten. He went to great length to point out that if there is harmony in a Yin-Yang balance, then foods can do much to make one look and feel youthful. In modern parlance, we may consider a Yin-Yang balance to be an enzymatic balance. This is brought about by eating in a proper and healthful manner.

Emperor Shen-Nung then offered his "secrets" of self-rejuvenation. When we read that many of these ancients lived well beyond the century mark during an era when medicines were nonexistent, we may well believe that the Emperor did tap the secret of perpetual life by creating an enzyme balance which he called a Yin-Yang balance so that foods could be assimilated. His "secrets" were jealously guarded until a copy of the scroll was made and other ruling monarchs made use of the knowledge.

Here are Emperor Shen-Nung's Five Secrets of Eternal Youth as offered in the forbidden and secret *Yellow Emperor's Book of Internal Medicine:*

1. *Eat only when hungry.* The ancients discovered that hunger is the voice of Nature saying that food is required. The enzyme benefit here is that when there is a feeling of natural hunger, these digestive elements create "appetite contractions in the stomach" which are Nature's way of saying they are ready to receive food. To overload the enzymes by habitual or compulsive eating is to upset the delicate Yin-Yang balance and invite ill health or inadequate absorption. The Oriental healers admonished their followers to favor the Yin-Yang elements (digestive enzymes) by eating only when there is a genuine feeling of hunger.

2. *Do not eat when in pain, when in mental or physical discomfort, or when feverish.* Ancient Orientals realized that pain, anger, fever, or inflammation each and all hinder the action of Yin-Yang, or interfere with the secretion of enzymes. Physical or mental upsets divert the enzymes away from the digestive organs. Food may become a solid mass in

the system. Whatever nutrients are present will be improperly assimilated. The Emperor Shen-Nung went to great lengths to point out that physical or mental distresses inhibit the flow of digestive juices and defeat the purpose of eating—to boost health and promote youth. He referred to it as disrupting the delicate Yin-Yang balance. He also said that the balance (enzyme secretions) can be influenced by the attitude. He said, "Joy is loved by Yin-Yang, while gloom is depressing to the Yin-Yang." During times of ill health or deleterious emotional attitude, food should be avoided since digestion will be impaired and enzymes cannot perform their health-building functions.

3. *Never eat until you have leisure to digest.* Emperor Shen-Nung learned that healthful digestion requires almost the entire attention of the enzymatic network. Blood is rushed to the digestive organs in large quantities. Enzymes bring about a dilation of the blood vessels in these organs to accommodate the extra supply of blood. Adequate digestion is difficult or impossible without a sufficient supply of blood-carrying enzymes. Emperor Shen-Nung wrote extensively about making time for the eating of food in order to have a happy Yin-Yang or enzyme balance. As the inventor of the plough, he noted wisely, "A full stomach does not like to think or plough." Together with a few of his carefully chosen aides who were sworn to secrecy, he urged for eating only when there was time for the Yin-Yang enzyme elements to create healthful digestion. He also suggested a rest before and after eating to help the Yin-Yang enzymes perform their rejuvenating tasks. Since the Emperor lived well beyond the century and a half span, we can understand why his secrets were so highly treasured.

4. *Do not drink with your meals.* Emperor Shen-Nung explained that the Yin-Yang balance should not be drowned before or with eating. He explained that the Yin-Yang enzyme elements become diluted if there is an outpouring of liquids with meals. He also urged thorough chewing to send forth a natural amount of liquids to the digestive system to create a natural moisture. The Emperor offered these guidelines:

- Drink your liquids an hour before an average meal.
- Drink your liquids thirty minutes after a fruit meal.
- Drink your liquids two hours after a starch meal.
- Drink your liquids three hours after a protein meal.

Emperor Shen-Nung explained that the Yin-Yang elements in foods themselves will improve digestion without being "drowned" by liquids from other sources.

He offered these suggestions that were said to "promise eternal youth-building Yin-Yang balance" for all who follow the secrets:

*Avoid cold drinks.* Any cold drink stops the action of the enzymes which must now wait until the stomach temperature has been raised to normal to resume digestive action. A cold drink will shock and chill digestive enzymes. To maintain a comfortable and living Yin-Yang youth-building balance, avoid cold drinks.

*Avoid hot drinks.* Any hot drink tends to weaken and enervate the stomach. Hot liquids often destroy the tone of the stomach tissues and weaken the *Yin-Yang* enzymes that must act upon food. Habitual hot drinks may often so weaken the stomach tissues that there is a prolapsus (sagging) of the abdominal muscles. To help favor the Yin-Yang balance, avoid hot drinks.

Emperor Shen-Nung urged for avoiding temperature extremes in any food, whether liquid or solid. He suggested a comfortable balance that was soothing to the Yin-Yang balance itself.

5. *Thoroughly chew all food before swallowing.* This guarded secret may well be the most valuable of all. Many folks bolt down their foods. The ingested mass often feels like a lump in the stomach. Whatever nutrients are present in the food may never reach the assimilation networks and there is an internal deterioration, despite a large intake of food. Emperor Shen-Nung always chewed food thoroughly and leisurely, since he said that this was the way to "flatter the Yin-Yang" enzymes in the digestive system. Food that has been thoroughly chewed is a prime source of youth-building substances when introduced to the digestive system. Emperor Shen-Nung explained that a "favorable Yin-Yang balance

begins in the mouth." It was his way of saying that digestion begins with chewing. Good chewing adds up to better digestion, better enzymatic assimilation of youth-building vitamins, minerals, amino acids and other essential nutrients.

These are simple rules, yet they are the core of helping to boost your enzyme functions and promoting youthful health. Emperor Shen-Nung and his close aides followed these Five Secrets of Eternal Youth and were able to enjoy a favorable Yin-Yang balance that gave them long and youthful lives, well beyond the century and a half span.

**How Ralph M. Regenerated His System with Emperor Shen-Nung's Five Secrets.** As a construction engineer, Ralph M. was faced with so many deadlines and so much travelling that he hardly had time to eat a good meal. He gulped down whatever food he ate, and could hardly wait to bolt down his food and go back to the field for construction supervision and related tensions.

Years of improper living gave Ralph M. serious stomach pains; he developed gas attacks. He could actually feel "bubbles" rising up in his digestive system after most meals. He slumped over. His muscles were weak. His thoughts were sluggish. He looked much older than his 37 years. A brief vacation did little. He still continued his habitual bolting down of his food.

It was while he worked on a research project concerning an oriental motif that he came across a report on Emperor Shen-Nung's immortal Five Secrets. He decided to try them. He adjusted his eating programs, taking longer hours for mealtime. He selected more natural and wholesome foods. It took three weeks of these simple yet effective enzyme soothing programs before he straightened up, had the color of youth in his previously sallow skin, and was alert. Most important, his stomach pains vanished with his gaseous attacks.

**Secret of Natural Healing.** Starches and sugars that are bolted down with water will ferment and give rise to acids. The enzymes are not given time to cause natural digestion. Therefore the starches and sugars that are improperly insalivated are unable to be fully digested. Enzymes are either unable to work on the foods or "drowned" in liquids gulped down at the same time. This caused the stomach churning and gaseous "bubbles" that created discomfort. Enzymes struggled with the bolted food and conflict was created. But when Ralph M. took time to chew his foods, letting salivary enzymes begin digestion, when he learned to eat only when in a comfortable mood and when hungry, he created a healthful Yin-Yang enzymatic balance that promoted better assimilation of nutrients–and gave him the look and feel of youth that food can create, if properly enzymatically metabolized. The ancient Emperor Shen-Nung knew of this process. It gave him a new life. It gave Ralph M. a new life too.

### *The Macrobiotic Way to Youthful Enzymatic Rejuvenation*

Macrobiotics is an oriental method of healing that is based upon principles laid down some 4000 years ago by Japanese healers. It was rediscovered at the turn of this century by Dr. Sagen Isiduka who reportedly helped cure thousands of people who followed the basic principles. His teachings were taken up by Dr. Georges Ohsawa, who reportedly was able to create longevity and rejuvenation in his patients by this all-natural program. Dr. Ohsawa writes[1] that he could cure his patients by this simple and recently revived ancient healing program. It calls for enabling the body's own digestive enzymes to create healing powers. He, too, explains that a Yin-Yang balance can be achieved by carefully controlled fasting programs. He says that some foods favor the Yin (passive, centripetal, feminine) principle, and other foods favor the Yang (positive, centrifugal, masculine) principle.

Actually, he explains that these two extremes are passive enzymes and positive or more aggressive enzymes. Mild foods are favored by the passive enzymes. Strong foods are favored by the positive enzymes. Youthful health is possible through a good equilibrium between these two opposing factors. A proper balance between foods gives these enzyme opposites an opportunity to function adequately.

**Controlled Macrobiotic Fasting Revives and Strengthens Enzymes.** Modern macrobiotic healers suggest dividing a program into a ten-day fasting schedule. Each day, devote the eating choices to specific foods. It is a form of the "mono-diet" in which one type of food is taken for one day with these benefits: (1) It

---

1. *The Philosophy of Oriental Medicine,* Dr. Georges Ohsawa, New York: Award Books, 1968.

enables opposing enzymes to rest for the day. (2) It encourages proper enzymes to perform their digestive work for that day. (3) This regenerates a digestive confusion or enzyme imbalance which Orientals call a Yin-Yang ratio.

A controlled macrobiotic fasting program helps revive the enzymes and cause a natural regeneration of the digestive organs. Here is the thousands-of-years-old secret of the ten-day "rejuvenation" with controlled macrobiotic fasting:

*First day:* Eat a wide variety of whole grains throughout the day. Grains may be cooked or raw. Chew thoroughly.

*Second day:* Eat 90 percent grains and ten percent vegetables. Take just a little water or liquid.

*Third day:* Eat 80 percent grains and 20 percent vegetables. Take just a little water or liquid.

*Fourth day:* Eat 70 percent grains and 20 percent vegetables. Now take ten percent soup. Take just a little water or liquid.

*Fifth day:* Eat 60 percent grains and 30 percent vegetables. Now take ten percent soup. Take just a little water or liquid.

*Sixth day:* Eat 50 percent grains and 30 percent vegetables. Now take ten percent soup. Take ten percent of any animal or dairy products. Take just a little water or liquid.

*Seventh day:* Eat 40 percent grains and 30 percent vegetables. Now take ten percent soup. Take 20 percent of any animal or dairy products. Take just a little water or liquid.

*Eighth day:* Eat 30 percent grains and 30 percent vegetables. Now take ten percent soup. Take 20 percent of any animal or dairy products. Take ten percent fruits and salads. Take just a little water or liquid.

*Ninth day:* Eat 20 percent grains and 30 percent vegetables. Now take ten percent soup. Take 25 percent of any animal or dairy products. Take ten percent fruits and salads. Take five percent desserts. Take just a little water or liquid.

*Tenth day:* Eat ten percent grains and 30 percent vegetables. Now take ten percent soup. Take 30 percent of any animal or dairy products. Take 15 percent fruits and salads. Take five percent desserts. Take just a little water or liquid.

*Reported Healing:* Orientals report that those troubled with digestive disorder, ulcer, gastritis, constipation, colitis, varying forms of indigestion, diarrhea, bloating, and other related problems were healed when going on a ten-day rejuvenation plan with controlled macrobiotic fasting. As you can see from the preceding program, it enables the Yin-Yang enzyme system to recuperate from years of erroneous mistreatment. On the first day, with just whole grains, your Yin-Yang enzyme system is able to rest and recover. Then the gradual introduction of vegetables, soup, animal and dairy products, fruits, soup and desserts, helps to gradually restore the health of the enzyme system. At the end of the ten-day program, the enzyme balance should be restored and a feeling of youthful health then follows.

*To Get the Most Enzyme Benefit Out of Your Macrobiotic Plan:*

1. Foods should be wholesome and natural. Avoid chemically grown or processed, adulterated foods.

2. Foods should be seasonal, preferably grown in your area so they are good sources of natural enzymes and nutrients.

3. Avoid commercially prepared spices or seasonings. Use sea salt or kelp (sold at most health stores and large supermarkets) or any vegetized salt substitute. Herbs are good too.

4. For beverages, avoid coffee and commercial tea. Use herbal teas. Avoid cola drinks.

5. Do not use foods made with baking soda as this is a chemical that is destructive to enzymes in the gastric system. Select yeast-made bread goods.

6. Chew your food very thoroughly, even if you are eating soft grains or luscious soft fruit. Salivary enzymes prepare this food for better digestion.

7. When on the program, Oriental healers suggest a slight reduction in the drinking of water. They explain that water is Yin or expansive and bloats the body. They suggest about eight ounces (one average glass) of liquid a day. They feel

that you receive enough liquid from food that you eat. *Note:* This macrobiotic secret enables your enzymes to invigorate and strengthen themselves to perform their work without interference. As the Japanese healers say, "It is like removing a crutch from the stomach so it can heal itself!"

For thousands of years, controlled macrobiotic fasting programs have reportedly rejuvenated followers by helping the enzyme networks revive-regenerate and send a stream of life-building nutrients throughout the system. This ancient oriental healer is just one of many other modern discoveries of the forever young programs of bygone days.

### *The Oriental Herb That Promises Eternal Youth*

For over two thousand years, Korean ginseng—an oriental herb—has been reverently used by the people in China, Korea, Japan, Southeast Asia and Siberian Russia whose combined populations are more than half the world's total. They have had unlimited faith in Korean ginseng for its vitalizing and restorative powers. These beliefs in the rejuvenating powers of ginseng have been held by all classes of oriental society. No royal household in the Orient was without ginseng. No aristocracy in Asia did not covet it as a mark of prestige and a source of self-rejuvenation. Of all medicinal herbs, ginseng was and still is considered to be the most effective. It is called the "rejuvenation herb." Its enthusiasts consider it the panacea of all diseases and age-creating debilities. Those who are sick will use ginseng to help restore youthful health. Those who are healthy will use it to build up resistance to aging and to make themselves stronger.

Today, after 2000 years, Korean ginseng enjoys the same high reputation among the Orientals. Its fame and acceptance is now spreading into European Russia, Germany, Switzerland, France and many other countries. It is gaining in acceptance in the United States. Many who have taken this "rejuvenation herb" have said that it gives a feeling of overall healing and restoration of health. They have said that ginseng gives them the look and feel of youth from the inside to the oustide, from head to toe. This most revered and worshipped "rejuvenation herb" is looked upon as the natural key to youth and health.

**The Youth Herb in Modern Science.** The botanical name for ginseng is *panax quinquefolia.* It is a member of the araliacea family and is a prime source of youth-building enzymes. Modern science has confirmed that the youth herb contains *amylase* and *phenolase.* These two herbs are minute but miraculous powerhouses of regeneration within the digestive system. These two ginseng enzymes are needed for the metabolization of certain amino acids as well as sugars and starches. They will use these nutrients to rebuild the internal cell-tissue network throughout the body; they then use the sugars and starches to provide a feeling of youthful and time-released sustained energy and vitality. The secret benefit here is that ginseng enzymes will provide the user with a steady and regulated amount of youthful buoyancy that is often manifested in sexual vigor. Many Orientals who are known for being virile in advanced age have credited ginseng with its valuable enzymes for giving such a boost to body functions.

**Where Ginseng Grows.** Korean ginseng grows in high and deep mountains of eastern Asia. It is harvested very carefully because it is sensitive to conditions such as clear air, proper humidity, suitable temperature, special soil quality and abundant water. *Special Benefit:* Because ginseng reacts adversely when treated with chemicals or artificial fertilizers, it is grown with a natural pesticide-free method. It is grown in rich soil with partial shade and watered regularly. Korean ginseng, that is labelled as such, is said to be grown organically and is the prime source of the youth-building elements.

*Most Effective Part:* The dried root is said to be the most effective part of this plant. To use, chew the hard and dry ginseng root. About five sliced pieces chewed daily is said to help send a stream of youth-building enzymes into the digestive system to give the person a feeling of youthful well-being and energy.

**How It Promotes Youthfulness.** Enzymes as well as vitamins and minerals in Korean ginseng root stimulate the central nervous system, alert the respiratory tract, rejuvenate the gastric and intestinal actions of motility, and block the hyperglycemic response to adrenaline. Enzymes take the saponins (viscous glucosides that are considered energy boosters) from the ginseng root and send them speedily into the bloodstream. Here these saponins have an exhilarating, tonifying, regenerating benefit to the entire body. Once the body's balance has been restored, there is a healthy desire to perform most youthful functions, including that of sex. Throughout the Orient, ginseng root is said to promote a youthful sexual urge among those who look upon it as a *natural aphrodisiac.* Many Oriental physicians prescribe this youth herb as a natural "sex tonic" from Nature. It is oriental reasoning that sex is created by Nature and that Nature should be used in helping to boost any weakness of this urge. Ginseng is an accepted *natural aphrodisiac* throughout the Orient. It is their secret of youth and health. It is now available in almost all health stores and herbal pharmacies throughout the United States. Ginseng root is said to be the most effective and is preferable over the other parts.

### *The "Youth Fruit" from the Pacific*

Throughout the Pacific islands of Hawaii, Australia, New Zealand, portions of Ceylon and India, the apple is looked upon as a "youth fruit." Legends abound in the youth-building properties of this simple but enzyme-rich fruit. It is often taken for granted. The apple is a treasure of those rejuvenating enzymes needed for better health.

**How Donald O. Used the "Youth Fruit" for Self-Rejuvenation.** As a travelling salesman, Donald O. found each day a little longer and more tiresome. His face became sallow and blotched. His eyes were bleary. His breathing was shallow. Donald O. felt his health slipping away bit by bit. His appetite was poor. His tongue was coated. Whatever he ate felt like a heavy weight in his easily upset digestive system. He thought he was "getting old before his time." His steadily slipping health made him take to his bed by late afternoon. He was sliding downward.

A salesman who covered a region populated by Asiatics told Donald O. that these people regarded the apple as a "youth fruit." They often ate several apples before each meal as a means of "rejuvenating" the digestive system so it could work at assimilating nutrients from foods to follow. The salesman gave Donald O. a bag of freshly washed apples and told him to try it.

Donald O. tried. The first two days he felt nothing. He was discouraged. But he continued, not wanting to offend his friend. By the fourth day he felt better. His complexion improved. His eyes were clear. He could breathe better. More important, he felt light, springy and youthful. What had happened? Here is what the "youth fruit" of the Pacific did for Donald:

- Enzymes in the apple sent a stream of pectin and malic acids into Donald's digestive system to absorb body toxins and tone up and aid the entire process of metabolism.
- Enzymes in this "youth fruit" then provided a feeling of bulk that pepped up the digestive system and improved intestinal mobility.
- Enzymes from the apple sent their own supplies of Vitamins A and C to help the body resist infection, keep cells and tissues healthy, assist in sluggish digestion, improve the eyesight, and strengthen the nervous system. These enzyme-propelled vitamins helped guard against gastrointestinal upsets and acted as natural "tranquilizers" and relaxers.
- Enzymes took apple pectin to help inhibit the absorption of cholesterol from the intestines into the bloodstream. It is the natural way to help control excessive amounts of cholesterol, that age-causing sludge that often causes premature aging.

*Oriental Recipe:* Donald O. followed this simple oriental recipe to rejuvenate his system. Mash apple into a very finely grated pulp, almost in the form of raw applesauce. Eat slowly with a spoon. *Secret Benefit:* Enzymes are known to be more effective if the apple pulp is grated very fine and the surface is exposed. Enzymes can then seize apple pectin more readily and send it shooting into the bloodstream. Here the pectin absorbs the cholesterol from the intestines more rapidly. This combination passes throughout the body without entering into the bloodstream. *This is a natural method that literally washes out the cholesterol from the system–thanks to apple enzymes.*

**Donald O. Bounces Back with Youth.** Daily eating of apples, either chewing or by using the preceding mashed Oriental Recipe, helped boost Donald's health. In due time, he became such a youthful and vigorous salesman that he was given a more lucrative territory. He outsold men who were half his age with half of his apple enzyme restored youthful energy. He owes everything to the apple–the "youth fruit" from the Orient.

### *The Golden Fruit of Eternal Youth*

History tells us that the first edible banana was found in the humid tropical region of Southern Asia. Ancient Chinese manuscripts referred to the youth-building secrets of the fruit. Authors of Greece, Rome and Arabia described it as the remarkable fruit of India. Artisans of Egypt and Assyria incised graceful pictures of bananas on their monuments. All extolled the youth-building benefits of the banana. Early explorers told of finding the plants in Hawaii and on the Easter Islands. A once-forbidden document of the Polynesians offers this "secret" of perpetual youth:

*Magic Youth Formula:* Mash bananas with honey, almonds and nut oil. Combine thoroughly. Eat regularly. It is said that this Magic Youth Formula contained secret properties that promoted a feeling of well-being and exhilaration from head to toe.

**How Banana Enzymes Offer Hope for Youthful Healing.** The secret here is that banana enzymes can help raise the blood sugar to a healthful level because of resistance to fermentation in the intestinal tract. Enzymes then bring about a youth-building metabolism of invert sugar.

Banana enzymes use pectins to create bulky residues that then create an alkalizing action on the digestive system to promote the slow release of natural carbohydrates. This creates a feeling of *gradual, time-released, soothing comfort.* A complete head-to-toe rejuvenation is possible when the starch-splitting enzymes in the digestive system go to work to convert food into usable nutrients for the body's use.

*Special Digestive Benefit:* Banana enzymes help make this fruit nonirritating and semisoft. This creates a soothing, coating action on the frequently tortured or inflamed and irritated digestive organs. This helps to "coat" the digestive organs to help relax and refresh them with nutrients that seep from the banana into the cells and tissues. Banana enzymes create a purine-free factor that increases the alkalinity of the urine and gives it greater ability to dissolve uric acid. Banana enzymes help favor retention of calcium and phosphorus to be used for tissue building. Banana enzymes create an alkaline residue that helps guard against the growth of putrefactive bacteria. Banana enzymes are able to suck up intestinal bacteria and slough it out of the system. Banana enzymes then help change bacterial flora and add bulk to intestinal contents. This helps rejuvenate the digestive process–the key to youthful health.

**Eat Bananas Daily.** The Orientals eat bananas almost daily: they do not wait until they are ill. Rather, they use bananas to prevent illness, which is rare enough, thanks to their awareness of health secrets from their revered and long-living, forever healthy ancestors. Now these secrets are available to you for your use and enjoyment.

*Highlights:*

1. Orientals believed the key to eternal youth was in a balance of Yin-Yang enzymes, made possible by Emperor Shen-Nung's Five Secrets Of Eternal Youth. Ralph M. regenerated his system with this secret formula.

2. The macrobiotic program of gradual enzyme rejuvenation calls for a simple ten-day controlled fasting program together with seven enzyme-boosting living habits. Easy to follow. Amazing in rejuvenation.

3. Ginseng is an oriental herb, available locally, that has a 2000-year-old history of restoring youth and virility to folks of advanced ages.

4. The apple with its treasure of enzymes is considered a "youth fruit" from the Pacific. It gave Donald O. a speedy bounce back to youthful health.

5. The banana is regarded a golden fruit of eternal youth. It is the oriental "natural medicine" with amazing enzymes with the ability to restore digestive-intestinal health and overall youthful response. The legendary Magic Youth Formula, using four foods found almost everywhere, reportedly exhilarates from head to toe.

# 6

## *How a One-Day "Enzyme Fasting" Program Helps Rejuvenate-Revitalize Your Body and Mind*

Since ancient times, fasting has been used as a means of rebuilding and restoring the body and the mind. In more recent days, fasting has been prescribed by physicians for those patients who seek to improve their digestive power and their enzymatic vigor, and promote a feeling of well-being of both body and mind. While some fasting programs call for several days of regulated eating as well as total abstinence from food, there are some programs calling for just one day of controlled fasting that reportedly will help relax-refresh-rejuvenate the enzyme system so it can function better in the process of assimilation.

### *How Fasting Helps Improve Youthful Health*

It is known that fasting will enable the enzyme system to perform *autolysis*, a recognized metabolic process by which the body's cells and tissues are dissolved or digested. Fasting enables the enzyme system to act upon stored-up wastes and toxins and then cast them out of the system.

Many persons experience a decline in health which can be traced to improper body nutrition and care. A slowdown of the vital endocrine system, a sluggishness of the metabolic rate, a decline of the process of assimilation are all symptoms of improper body care. Prolonged abuse of the body and a slowdown of the organs of elimination will lead to a residue of metabolic toxins and cellular wastes in the system. An accumulation of these clogged-up wastes may lead to a condition known as autointoxication. The body's natural resources try to self-cleanse and correct any disturbances, but the overwhelming accumulation of wastes and toxins clog up the smooth working efficiency of the internal viscera. There is a general slowdown of metabolism and the end result is internal toxemia. To help promote internal washing, fasting is a recognized method of natural regeneration.

Here are five benefits of simple but effective "one-day" enzyme fasting programs that will be outlined in this chapter:

1. *Autolysis.* A reduction of food intake compels enzymes to work upon the stored-up substances left in the body. Enzymes will then create the process of metabolism by which the body will digest the toxic wastes and metabolic sludge through the process of autolysis or self-loosening. Enzymes prompt the metabolic system to decompose and dissolve the diseased tissues and cells. Enzymes create the autolysis process to act upon sick secretions, sticky accumulations, corrosive wastes, and excess fatty deposits, especially cholesterol. Autolysis is prompted by enzymes to metabolize impure and unhealthy wastes that are consumed and then cast off.

2. *Better Respiration.* Metabolic wastes and cellular refuse often clog up the organs of respiration as well as those of elimination. This impedes normal respiration-elimination. During a controlled enzyme fasting program, the body miracle workers have no food to act upon and are liberated so they can work upon the accumulated wastes that have clogged the lungs, the kidneys, the liver, the venous-arterial network, the subcutaneous skin cells and pores. Enzymes seize these wastes and help to expel them through the normal channels of elimination. You can see the sloughing off of these wastes in the form of dark urine, excessive and often odorous perspiration, bad breath, and abundant waste discharge. It is Nature's way of cleansing the organs of respiration-elimination so they can be properly "aerated" and function more efficiently.

3. *Digestive Rejuvenation.* Daily intake of food calls for never-ending work on the part of the digestive system. The ancients understood this and decreed that there were to be religious fasts so that the digestive system could enjoy a rest, like the other parts of the body. While ancients labelled it a religious-spiritual obligation, today we know that fasting is a digestive obligation which enables the organs to rest, relax, and refresh. After a one-day controlled fast, the enzymes bounce back with renewed strength and vigor and help create better assimilation.

4. *Improves Emotional Health.* A one-day fasting program enables your enzymes to do more than just clean out your internal organs. It helps improve the process of metabolism. Furthermore, cleansed digestive organs are able to better assimilate nutrients that are sent via a sparkling bloodstream to all parts of your body for regeneration. Your brain, too, benefits from a clear bloodstream that is able to transport nutrients to the cells and tissues that influence emotional and mental health. Enzymes are intimately involved with brain function via feeding of the brain cells. A healthy body is better able to feed the brain. Enzymes are invigorated to perform this vital function, together with other benefits, after a refreshing one-day fast or day of rest.

5. *Youthful Vigor.* Fasting enables the enzymes to flush toxic wastes from the body. This now gives enzymes the renewed ability to convert new food into youth-building energy and vigor. A cleansed body welcomes nutrients that can work more efficiently with new enzyme ability. Youthful vigor is often the reward for a simple one-day fasting program.

Fasting has long been hailed as an all-natural internal cleanser. It enables enzymes to wash away accumulated wastes and toxic acids *without* interference of ingested foods which often *divert* work from autolysis to metabolic digestion. When the digestive system is left free from foods which require heavy digesting, then enzymes are free to turn attention to the accumulated residues and help create self-cleansing. It helps build and rebuild health from the inside to the outside.

### *How to Begin Your One-Day Fasting Program*

1. Prepare yourself in advance, mentally and physically. For several days before, cut down on heavy fats and starches. Boost your intake of fresh raw fruits and vegetables and their fresh juices. This helps alert your enzymes to the task ahead.

2. On the morning of your fast, begin with a glass of fresh raw fruit juice. Prune juice is popular because it begins the natural process of elimination. Throughout your day of fasting, drink herb teas. Flaxseed herb tea is often suggested because it contains ingredients that help promote body elimination of toxic waste substances.

3. You may continue with your work during your day of fast, but it is best if you try to eschew serious mental or physical work. Your body's metabolism should be conserved for the enzyme action upon waste material. If you perform arduous work, that energy is taken away from the enzymes: this may weaken their autolytic vigor. So prepare for reduced amounts of activities—but there is no need to take to your bed! Actually, keeping mildly active is favorable to better body enzyme metabolism.

4. Take a regular walk to help boost your respiratory system. Many fasters find that a one-hour walk in the early morning, then a one-hour walk at noontime and a final one-hour walk at evening is most refreshing and invigorating. Healthy fresh air and sunshine are beneficial.

5. Since increased casting off of wastes through skin, respiratory organs, and urine will indicate self-cleansing, it is essential to take a daily bath. Water should be comfortably warm. Use a rough wash cloth and "scrub" at your skin from top to bottom. This helps wash away the sloughed-off impurities and toxic wastes that might otherwise clog up your skin pores and defeat the purpose of body ventilation. For good results, take a bath at night, before retiring.

6. Enjoy a good night's sleep on a comfortable bed. Your bedroom should be comfortably warm. Open one window to permit the exchange of air. Keep out of drafts.

These few simple preparations will do much to help your enzymes cleanse your body during a one-day controlled fasting program.

### *The Mucusless Fasting Program*

An early proponent of enzyme fasting was Professor Arnold Ehret, who reported that many people could enjoy better health and extended youth by cleansing out mucus from the system. In brief, Professor Ehret explained[1] that foods are divided into two groups:

*Mucus-Forming:* All animal products including meat, fish, eggs, and dairy products.

*Non-Mucus Forming:* All fruits, vegetables, and grains.

Professor Ehret prepared a two-day simple fasting program for his followers. He said that the program may be followed during a two-day weekend—or one day at a time. By emphasizing the use of certain non-mucus forming foods, Professor Ehret stated that enzymes could then have the ability to devote attention to metabolizing the accumulated mucus and poisons and then cast them off. He stated that when mucus-forming foods are taken, enzymes must devote time to the digestion of such foods and are inhibited from cleansing. He suggested this special fasting program that eliminated mucus-forming foods and enabled enzymes to perform their valuable cleansing action upon the previous accumulation of internal mucus.

Professor Ehret suggests either a one-day or a weekend fast. Here are the various controlled fasting programs. he prepared:

**Fasting Program #1.** *Breakfast:* A fresh raw juice or herbal tea or coffee substitute. *Lunch:* A combination salad of grated carrots, two teaspoons of stewed spinach, sliced tomato and oil and lemon dressing; a baked cauliflower. *Dinner:* Stewed prunes

1. Professor Arnold Ehret, *Mucusless-Diet Healing System,* Beaumont, California: Ehret Publishing Co., 1971.

and apricots with cottage cheese. *Liquids:* Fluids should not be taken with the meal. Allow at least 30 minutes to pass before taking liquid after eating.

**Fasting Program #2.** *Breakfast:* None. *Lunch:* A combination salad of cole slaw, two teaspoons of cooked string beans, sliced cucumbers and a dressing of oil and lemon; a baked squash. *Dinner:* Applesauce with stewed peaches and cottage cheese. *Liquids:* Only to be taken at least 30 minutes after the meal is finished.

**Fasting Program #3.** *Breakfast:* None. *Lunch:* Baked apple. Allow ten to 15 minutes to pass. Then have a combination salad of grated carrots, two teaspoons of stewed spinach, and sliced cucumber. For dressing, use juice from spinach or baked apple. *Dinner:* Baked cauliflower with a salad of lettuce and cucumber.

**Fasting Program #4.** *Breakfast:* None. *Lunch:* Stewed apricots. Allow ten to 15 minutes to pass. Then have a combination salad of cole slaw, cooked string beans and tomatoes, using the juice from the string beans or apricots as a dressing. *Dinner:* Baked broccoli, followed by celery and cucumber. *Liquids:* After the meal.

**Fasting Program #5.** *Breakfast:* None. *Lunch:* In the summer, use just one kind of fruit such as peaches. In the winter, use raisins and apples, followed by a little lettuce. *Dinner:* Grated carrots, cooked green peas and sliced tomatoes followed by baked squash. *Liquids:* No change. Drink after and not during the meals.

**Fasting Program #6.** *Breakfast:* None. *Lunch:* In the summer, use just one fruit such as cantaloupe. In the winter, use figs and oranges; a few nuts may also be added. *Dinner:* Cole slaw, cooked spinach and sliced cucumber, followed by baked cauliflower. *Liquids:* To be taken after the meal.

**Fasting Program #7.** *Breakfast:* None. *Lunch:* In the summer, use just one kind of fruit such as apricots. In the winter, use raisins and oranges; a few nuts may be added. *Dinner:* Fresh or stewed peaches followed by cooked squash. *Liquids:* To be taken after the meal.

**Fasting Program #8.** *Breakfast:* None. *Lunch:* In the summer, use just one fruit such as peaches. In the winter, use figs and apples; a few nuts may also be added. *Dinner:* Fresh or stewed apricots followed by a combination salad of cucumber, grated carrots and lettuce.

*Extra Tips:* Professor Ehret says the preceding controlled fasting program enables the enzymes to slough off accumulated cells and may also attack the fat cells, creating a desired weight loss. But if you do not want to lose weight, he suggests eating a baked potato or toasted wheat or rye bread at the end of each meal. If you have an intense craving for meat, then eat only vegetables and no fruits as a means of enabling your enzymes to continue their washing action and providing you with a feeling of satiety. (He explains that when enzymes digest vegetables alone, they use the natural carbohydrates for providing a feeling of fullness and contentment that eases the urge for meat-eating.)

*Most Effective Enzyme Fasting Plan:* Professor Ehret believes that a mono-diet is most pleasing and beneficial. That is, *just one food in season is eaten.* The benefit here is that just the particular fruit or vegetable enzyme is used and the others are free to metabolize stored-up body wastes and fat. He suggests a lunch of ripe bananas, or a dinner of a large bunch of grapes. This spares the body's supply of fat and mucus-melting enzymes (since the body is now spared intake of fat and mucus-causing foods) to be able to metabolize the stored-up fat and mucus.

### *The Ehret Program for Better Enzyme Power in Fasting*

1. Always eat your fruit first so that enzymes can then be propelled throughout the system for their mucus-melting work.

2. Liquids should not be taken with a meal. Wait at least 30 minutes after eating before you drink. Liquids often dilute and weaken the power of digestive enzymes.

3. Fruits, vegetables and herbs are man's most natural foods. They supply the body with vigorous enzymes that promote the most favorable self-loosening of toxic wastes.

4. The mono-diet is enzyme favoring. Limit the variety of foods you eat at one meal. This also puts a stopgap on your appetite and helps bring about reducing.

5. Because nuts are sources of essential fatty acids, they should be eaten at the end of a meal. Do not eat nuts with

juicy fruits because the enzymes are in conflict with one another at the same time. Self-cleansing may thus be slowed down.

6. Eat in moderation. An excess of food overworks the enzymes and torments the digestive processes.

7. Foods should be natural, wholesome and organic. Chemicals add more toxic wastes and corrosive acids to the system and interfere with enzymatic healing.

Many of Professor Ehret's followers were able to "bounce back" with youthful health by a regular one-, two- or even eight-day mucusless fasting program. The joy of this program is that it may be followed for just one day or as often as you desire. It reportedly is helpful in just one day. Of course, if followed regularly, the program helps supercharge your enzymes to continue their work in mucus-cleansing to help give you a feeling of youthful health.

**How Phyllis R. Enjoyed Renewed Youth and Freedom from Allergies on the Easy Ehret Fasting Program.** Years of cold-catching and recurring asthma, as well as embarrassing sensitivity to the slightest dust, made Phyllis R. look much older than her years. A problem here was that as her enervated body fought back against the invasion of allergy and cold germs, it left her wasted and nutritionally deprived. A major amount of her vitamins, minerals and enzymes had to be used up to defend her against allergic attacks. This left her body deprived so that she had cold hands and feet, reportedly had forms of anemia, looked pale, and was always shivering even in warmer weather. Phyllis R. had tried many different suggested healings but they were temporary. She heard about the traditional Ehret fasting program and followed it for just one day.

**Feels Allergic Spasms Easing Up.** At the end of the day, her sniffles subsided and she felt warmer. She looked a little more rested. The Ehret program had sent a stream of enzymes to her respiratory tract; specifically, her bronchial tubes and diaphragm and her intercostal muscles were cleansed of some of the accumulated mucus. She had a history of eating heavy fats and dairy products. (Metabolized milk often leaves curds on the body organs and this impedes the free flow of enzymatic action.) But with the use of raw enzymes from non-mucus forming foods in the Ehret program, much of the mass could be sloughed off.

Breathing better, Phyllis R. went back to her regular heavy fatty eating and soon developed more serious allergic complications. But when she followed the Ehret fasting program for just three days, she bounced back with vigor and stamina. She was soon healed of her allergies, had warm hands and feet, and looked sparkling and young with color in her cheeks. She fasts one day a week, selecting any of the preceding Ehret fasting programs, and follows his basic rules of more fruits and vegetables and fewer heavy, fatty meat and dairy meals. Now she looks young and has the energy of a youngster too.

*How Enzymes Cleansed Mucus:* In problems of allergies, the bronchial tubes are often clogged with impurities and mucus. A cough tries to expel the sludge. It is Nature's way of suggesting the use of enzymes to cleanse these bronchial tubes. Mucus accumulates on the elastic bags of the lungs, the diaphragm and intercostal muscles which span the gaps between the ribs. Air is inhaled but has difficulty in going down the diaphragm and connected thorax. The muscles lining this dome-shaped network become clogged with mucus and deprived of oxygen. The allergic person chokes and sputters. But self-cleansing by the use of enzymes works to send mucus-dissolving elements to the accumulated toxic wastes in the bronchial region. By loosening and washing away these wastes, the enzymes help cleanse the deposits and now enable the lungs to breathe better. A freer passage of oxygen is the first step toward healing of allergic disorders. A one-day fast (or several days of the Ehret program for more benefits) is the natural way to clean out mucus from the lungs—the lifeline of youth and health.

### *How a Simple One-Day Vegetable Fast Rejuvenates the Body*

Earl S. works under typical business executive tension. He cannot (or will not!) reduce his work load. He is overly conscientious, even under the risk of a threat to his health. His basic problem is that of *acidosis,* an alteration of the acid-alkaline balance of the bloodstream. It produces churning and ulcer-like wrenching in the pit of his stomach. He feels all knotted up; often, he feels like doubling over at the waist. This has taken its toll on his appearance as well. He walks stooped over, looks waxen and pale, has bleary eyes, and often has the shakes.

Not wanting to lose his job to a younger man (he is only 34 but looks older because of the acid problem), he goes on a periodic one-day vegetable fast. He learned about it when he took a vacation at a health spa where fasting was part of the healing program. Earl S. finds that just one day a week on raw vegetables works miracles in rejuvenating his body.

*Simple Program:* Just one day a week, everything that is eaten (solids and liquids) should be from a vegetable source. *Raw* vegetables and *raw* vegetable juices are prime sources of top-notch enzymes. *Breakfast* could be a raw vegetable drink, with a plate of raw vegetables. *Lunch* could be another raw vegetable drink and a plate of steamed vegetables together with a side dish of raw vegetables. *Dinner* could be a plate of raw vegetables with a dressing of honey and vegetable juices, a dish of steamed or baked vegetables, a bowl of vegetable soup with natural brown rice, and a cup of herb tea.

*Benefits:* Earl S. finds that this program helps ease his burning-churning problem of acidosis. He is able to walk straight because the knot in his stomach has become untied. He has better coloring, brighter eyes, and a more alert mind. He looks, feels and acts younger.

*Enzyme Cleansing:* Vegetable enzymes work without interference of ingested fats and sugars. They stimulate the production of alkaline substances taken from the metabolized vegetables. The enzymes use these alkalies for the combustion and elimination of corrosive acid residues. Enzymes then help reach an equilibrium between acids and alkalies to restore a soothing gentle reaction to the digestive system. This reflects in a more youthful appearance as well as a more youthful emotional outlook. Vegetable enzymes hold digestive alkalinity more constant by maintaining a balance of dissolved carbon dioxide (acid) and sodium bicarbonate (alkali). These are *natural* body compounds created by metabolism. Alkaline enzymes from the raw vegetables help eliminate excess carbon dioxide, promote better combustion of fat (often poor combustion leads to acidosis), and restore a proper balance in the system.

By helping to remove acid toxins from the system, the raw

vegetable enzymes work freely and totally upon restoring alkaline health to the burning digestive system. A simple yet effective program, according to the natural healing of one-day fasting.

### *How a One-Day Raw Fruit Enzyme Fast Helps Reduce Blood Cholesterol*

You are as young as your arteries. An age-causing problem is that of cholesterol. This is a fatty substance found in most animal fats and dairy foods. Your body makes cholesterol and can sustain you with a sufficient supply to keep you healthy. But an excessive intake of animal fats and dairy foods may give you an excess of cholesterol.

Cholesterol becomes deposited throughout the body, narrowing and clogging the arteries. A cluster of blood cells may form a clot large enough to shut off the flow of blood. This may precipitate a problem of arteriosclerosis (hardening of the arteries) or a more serious heart attack. The arteries become impeded and are considered "old," even in young persons. Truly, you are as young as your arteries, and cholesterol may make them (and you) old before your time.

**The Enzyme Cholesterol-Melting Factor in Raw Fruits.** Fresh raw fruits and their juices are virtually free of cholesterol. They are often prescribed as a "perfectly safe" food for folks on a low-cholesterol program

Specifically, the enzymes in raw fruits and their juices create a natural "micro-electric" cleansing action in the cell tissues of the arterio-neural network. This boosts cell respiration and stimulates cell metabolism. The fruit enzymes boost the cell's power of resistance by cleansing it of the clumps of cholesterol; this brings about better cellular renewal since vitamins and minerals now can nourish the tissues. Metabolism is then able to proceed more youthfully.

**A Simple Program Helps Housewife.** Doris U. loves fatty foods. As a dedicated housewife, she spends much time in the kitchen, sampling her fatty rich foods in addition to eating heartily at mealtime. She has a higher cholesterol count than is suggested. She feels she cannot control her urge to eat fatty foods. So she goes on a one- or two-day all-fruit fast. She is able to resist the urge by eating lots of fresh fruits and drinking lots of fresh raw juices. Reportedly, this helps *control* her cholesterol deposits and she does look much healthier with clearer skin and more vitality. But she has to go back to the fatty foods and this often slows up her energies. (Fatty foods require more digestive action and enzymes are rushed to the stomach in great amounts so that other bodily functions slow up.) But Doris U. does control her cholesterol with this easy program:

*Breakfast:* Fresh raw seasonal fruit salad with a dressing of honey and lemon juice. A large glass of a fresh fruit juice. *Lunch:* A bowl of fresh fruit with desired fruit juice dressing; a dish of cooked fruits to ease the urge for eating warm foods. A glass of fresh fruit juice. *Dinner:* Stewed fruit in their own natural juices; a fresh fruit bowl with chunks of juicy fruits; a glass of steamed apple juice with a sprinkle of cinnamon.

*Benefits:* In the absence of food cholesterol, Doris' own enzymes are now free to digest and slough off the accumulated cholesterol in her body. Fruit enzymes also work to create this enzymatic flow and it does help control the cholesterol deposits. It is a form of cellular washing. She calls it her "fruit medicine" because it makes her feel good and look good as well. It also helps control her runaway appetite and she has hope for soon being able to cut down on the excessive intake of fatty foods—thanks to the satiety values of fruit enzymes.

## How to End Your Fast

Ending a fast the right way is important if you are to enjoy all the benefits of the fast itself. To end your fast, follow these simple suggestions: (1) Take several days of slow transition to your regular diet. (2) Chew all food very thoroughly before swallowing. (3) Do not overeat. Gradually make these adjustments:

*First Day After Fast:* Select warm foods that are either fruits, vegetables or grains. Broths and soups are soothing. Whole grain breads are welcome to the digestive system.

*Second Day After Fast:* Begin your intake of fermented milk in the form of yogurt (made from skim milk) or buttermilk; add vegetable broths or purees. Take no salt or sharp spices. Include cooked vegetables throughout the day as well as some cooked fruits. Fresh juices are soothing. Warm herbal teas or coffee substitutes are good.

*Third Day After Fast:* You may now start to take broiled meats, poultry, and some fish. Be sure to have a bowl of fresh fruit or vegetables regularly to give your body all the vital enzyme elements needed to continue health restoration.

*Above all:* Do not overeat; do not overload or overstrain. Eat wholesome, natural foods and give your enzymes the raw working materials from which new life and new health are made.

One day fasting programs can become part of your health picture of the future. They can keep you looking and feeling young through cellular respiration and regeneration—the key to a "forever young" body and mind.

### *Special Points:*

1. Fasting is an ancient and currently popular way to enable enzymes to create autolysis, a metabolic process of cellular rejuvenation. Some of the benefits of fasting include better respiration, digestive rejuvenation, improved emotional health and youthful vigor.

2. To begin your one-day fasting programs, just follow the amazingly easy six steps as outlined. They form part of the enzyme-regeneration picture of youthful health.

3. Professor Arnold Ehret offers you eight different one-day fasting programs to help cleanse out the mucus from the system. The famed and reportedly successful Mucusless Fasting Program emphasizes the use of enzymes to slough off accumulated wastes and toxic crystals.

4. To help boost enzyme action on mucus, follow the seven-step Ehret program.

5. Phyllis R. bounced back to health and enjoyed freedom from allergies by following the easy Ehret Fasting Program.

6. Earl S. soothes acidosis the natural way with a simple one-day raw vegetable fast. It makes him look and feel younger than his competitors.

7. Doris U. finds that a raw fruit enzyme fast is the natural way to help control cholesterol and keep her looking much younger.

# 7

## *How the Swiss Use Raw Grain Foods with "Youth Enzyme X" for Renewed Youth*

The hardy, long-living, rugged Swiss people have discovered that raw grain foods contain a powerful substance that reportedly helps give them a feeling of renewed youth.

Many Swiss healers and physicians will prescribe the eating of different raw grains to help correct digestive disorders, create a feeling of overall rejuvenation and help restore more youthful vitality to persons of all ages.

The Swiss emphasize *raw grain foods* as opposed to the lifeless processed or precooked cereal grains. The Swiss report that *raw grains* are prime sources of a Nature-created enzyme that helps spark up body functions and create miracle-like youthful regeneration. The eating of fresh raw grains is often prescribed by Swiss physicians for those folks who need a natural "medicine" from Nature. The reported benefits are so satisfactory that Swiss healers have labeled the rejuvenating nutrient as "Youth Enzyme X." It is found *only* in raw grain foods. It is *not* found in preserved, processed or chemically treated precooked cereal grains. The processing destroys the highly sensitive enzymes and the finished product is regarded as "lifeless," incapable of creating new life as is the miracle of *raw grain foods* with "Youth Enzyme X."

**"Youth Enzyme X" in Raw Grains.** Raw grains are Nature-created seeds. Sealed within the kernel are a treasure of substances that Nature has placed there to provide nourishment and eternal life for the plant that will emerge. Enzymes in raw grains are so powerful that they can sustain the seed for decades upon decades with a miracle power of self-rejuvenation and self-renewal. Nature-created raw grains can remain fertile for an indefinite time span, thanks to the nourishment created by "Youth Enzyme X" placed within the tiny kernel. This same "Youth Enzyme X" is reportedly rejuvenating to those who eat raw grains, and the hardy Swiss have enjoyed better health and youthful vigor because of this discovery.

Raw grains appear to have been given special favor by Nature in that the whole seed contains practically all the elements required by the body for sustenance. These elements help keep the grain seeds alive and fertile and resistant to the ravages of illness. They are said to help the Swiss enjoy the same "forever young" benefits. "Youth Enzyme X" is found solely in raw grains and those that are germinated; that is, raw grains that contain the "germ" or "seed" of fertile life. The Swiss emphasize the use of nonprocessed, natural raw grains because these have been found to be most effective and most potent with "Youth Enzyme X" for better health. Such natural raw grains are available at most organic food outlets and health stores.

### *How to Make Your Own "Youth Enzyme X" Germinated Seed*

The Bircher-Benner Sanatorium of Switzerland prescribes this special easy-to-make "Youth Enzyme X" Germinated Seed recipe that reportedly will provide an all-round strengthening effect and a feeling of perpetual youth.

*Enzyme Rich Seeds:* raw wheat, rye, oats, barley. Use whole grains only. Organic seeds are prime sources of "Youth Enzyme X."

*First day:* In the evening, wash the organic grains in a sieve under running water and put in a bowl. Cover with water. Keep at room temperature near stove or radiator.

*Second day:* In the morning, rinse and spread out to dry on a flat plate. Keep at room temperature near the stove or radiator.

*Third day:* In the morning, rinse and spread out to dry on a plate. In the evening, place in a bowl and cover with water. Keep at room temperature near the stove or radiator. The grains should by now have developed one-half inch sprouts. The "Youth Enzyme X" is now at its peak.

*How to Use:* You may make a mixture of these different germinated seeds, add some fruit juice in a bowl, sprinkle with raw wheat germ and use as a powerhouse breakfast. Or you may moisten with skim milk and add sliced fruit with some raw wheat germ for a delicious and youth-building breakfast. The Swiss recommended mixing the germinated "Youth Enzyme X" seeds with bananas, honey and water for a tremendous boost of body and mind. It is prescribed for those who want to bounce back with better health and vigor at all ages.

### *Raw Grains That Are Nature's Sources of Powerful Enzymes*

Among some of the highly recommended raw grains that the Swiss use for powerful enzymatic rejuvenation are these:

**Oats.** Enzymes in oats take out valuable protein and essential fatty acids from the seeds and introduce them into the bloodstream for self-rejuvenation. Oat enzymes also act as silicon carriers that enable the body to resist congestions and the formation of toxins and acid reaction. Oat enzymes also contain such blood-building and health-restoring minerals as phosphorus, potassium and magnesium.

**Rye.** The "Youth Enzyme X" factor is most powerful in rye berries which take up valuable Vitamin E to promote better heart health, and the precious essential fatty acids for transport of Vitamins A and D to promote a healthier skin and better bone structure. Rye enzymes use these vitamins for the building and rebuilding of the body structure and its various processes.

**Corn.** Enzymes in corn kernels are prime sources of essential fatty acids that transport natural minerals throughout the bloodstream. The outer covering of corn kernels contain the minerals, and the second covering contains a phosphorus compound. The seed germ itself is a treasure of a vital nuclein which is used by the enzyme to help maintain health, strength and vitality. "Youth Enzyme X" is found solely in organic and raw corn kernels. The so-called "denatured" corn is void of the important organic nutrients as well as the precious enzyme and can be considered "lifeless" food that fills up and fattens but does not rejuvenate. Swiss healers prescribe untreated natural raw corn kernels as a prime source of "Youth Enzyme X."

**Barley.** Here is a popular Swiss grain that is often mixed with other raw grains so that its original mild vitamin-mineral supply is complemented by the more vigorous nutrients of other seeds. "Youth Enzyme X" takes up the essential fatty acids of barley and uses them to lubricate its cellulose fibers so that a natural bulk is created to enter in the natural process of elimination. *Note:* Many Swiss physicians will prescribe raw barley seeds to be mixed with fruit juice, milk or water, together with some honey for flavoring as a *natural laxative.* Swiss sanatoriums use this simple all-natural mixture for constipated patients.

**Millet.** This is an Old World favorite that was used for thousands of years. The "Youth Enzyme X" in millet acts upon its protein content to take out ten of the basic amino acids and then combine them with ingested foods to create a harmonious reaction in the system. "Youth Enzyme X" in millet is used to create a non-acid-forming reaction that often helps ease problems of acidosis. A unique benefit of "Youth Enzyme X" is that it predigests millet and makes this grain easily digestible for many folks of all ages. It is believed to bolster weak digestive functions in the elderly, giving them "younger" metabolism.

**Buckwheat.** Here is a heavy raw grain with a rich nutty flavor. The whole buckwheat berries are often called "groats" and are the famous "kasha" that is so filling and can serve as a powerhouse of energy, strength and vitality for the day ahead if eaten at breakfast.

**Bran.** The outer coarse coat of grains, it is a rich source of many minerals. Its built-in "Youth Enzyme X" has a natural regulatory effect in the digestive system. This enzyme in the bran absorbs moisture and thus keeps the liquid in the intestinal canal. By this enzyme-regulated process, the intestinal duct is kept soft, ductile, elastic and youthful. Bran enzymes stimulate the peristalsis and increase the waves of rhythmic contractions and expansions so necessary for the transportation and elimination of the contents of the bowels. In the colon (large intestine), bran enzymes deliver needed vitamins and minerals to be sent to the bloodstream which will use these enzyme-delivered nutrients for the nourishment and sustenance of billions of body cells. Bran enzymes metabolize such minerals as iron, iodine, potassium, magnesium, phosphorus and many proteins, together with lecithin and phosphates. All of these nutrients are inert within raw bran; when raw bran is eaten, the "Youth Enzyme X" in the seed is alerted to perform its metabolic activities. Bran has long been considered a miracle of youth in a seed, thanks to the living enzyme within its coat.

**Raw Wheat Germ.** This is the germ or the embryo of the wheat kernel and is used by the Swiss for its rich endowment of vitamins, minerals and grain proteins. "Youth Enzyme X" in raw wheat germ takes out Vitamin E to be used as an arterio-neural washing agent and also to prevent fats from becoming rancid within the system. Raw wheat germ is a prime source of enzymes that offer a form of grain protein which is reportedly the best in nonmeat foods. Wheat enzymes also use iron for blood building and hemoglobin formation. Raw wheat germ is the embryo of the entire wheat kernel which is a rich source of this "Youth Enzyme X" that reportedly promotes better health, stamina and vigor for the hardy Swiss.

*Additional Benefits of Raw Grains:* The chewy texture of raw grains alerts sluggish taste buds and boosts a healthy appetite. "Youth Enzyme X" in raw grains improves the taste for wholesome and good food. Raw grains are free of additives and artificial colorings as well as synthetic flavors. Swiss healers recommend the use of raw grains for folks on limited salt or sugar programs. Packaged cereals are laden with such harmful ingredients. The raw grains contain natural flavoring agents which are used by "Youth Enzyme X" for promotion of better health.

**Raw Grains Boost Digestive Vigor.** "Youth Enzyme X" promotes a natural boost to digestive processes. This enzyme macerates nutrients from the grains which reach the digestive system. Then the enzyme draws out essential qualities which are sent speeding throughout the bloodstream to create a natural digestive rejuvenation. The use of raw grains for youthful invigoration is part of the Swiss heritage and their secret for perpetual youth.

### *How a Simple Raw Grain Breakfast Provides Day-Long Vitality*

The raw grain breakfast known as Swiss Muesli is believed to have been born in this mountainous region. Needing a breakfast that would provide the hardy Swiss with body warmth, youthful strength and day-long vitality, they created this all-natural combination of raw grains and food. It is reported that Swiss Muesli helps create strong teeth and bones as well as a feeling of youthful vigor.

**Make Your Own Swiss Muesli.** Use natural, organic foods for best results. These ingredients make one portion.

1 tablespoon rolled oats, soaked previously for 12 hours in 3 tablespoons of water
1 tablespoon lemon juice
1 tablespoon yogurt
1 large apple (or 2-3 small ones)
1 tablespoon grated filberts or almonds

Mix lemon juice and yogurt to a smooth cream. Add to oats, stirring constantly and thoroughly. Now grate the washed apples (remove tops, cores and blemishes) into the mixture, stirring frequently to prevent discoloring. Sprinkle nuts over the finished dish and eat promptly. (Note: Add 2 tablespoons of fresh fruit juice for added flavor.)

**How "Youth Enzyme X" in Swiss Muesli Promotes Vitality.** Gloria G. was the victim of "mid-morning slump." No matter what she ate for breakfast, she would experience a letdown by mid-morning. Her thoughts were blurred. Her attitude was contrary because of her inability to act coherently. Her "lifeless" breakfast invariably consisted of sugar-sweetened, chocolate-covered processed cereal, a can of syrupy fruit slices and coffee. This gave her "lifeless" nutrients that had been destroyed by the processing and canning. She experienced a brief spurt of energy that then plummeted downward to create a reverse reaction of complete lethargy. She acted much older than her 35 years.

When she was told to eat "living" foods for "living" health, she tried the Swiss Muesli breakfast. Here is what this raw grain breakfast did for Gloria G. Its natural "Youth Enzyme X" took the carefully balanced Vitamin B complex and Vitamin E supply from the grains to be metabolized into better vigor and more youthful strength. Only whole grains contain this twin balance of these two vitamins. Processing separates the nutrients from the grain and reduces as well as upsets the energy-boosting B and E vitamin ratios. But in the Swiss Muesli whole grain breakfast, the abundant balance of these vitamins was used by the "Youth Enzyme X" to be transformed into living health and vitality.

Gloria G. ate this raw grain breakfast for over a week. She soon felt so alert, active and energetic by mid-morning that there was no stopping her. Thanks to "Youth Enzyme X," with its ability to use the nutrients in raw grains for energy-boosting, Gloria G. now looks young, thinks clearly, and is no longer given to stubborn confusion.

### *The Swiss "Raw Grain" Secret of "Instant Energy"*

On a vacation trip to Switzerland, Evelyn I. felt quite refreshed. She was energetic to the point where she joined a small group of mountain climbers on a guided tour up the famous Swiss Alps. The tourists soon became tired. The guide was just the opposite. He had limitless energy for a man in his late sixties. Whenever they had a "rest stop," the tourists would munch candy bars, sweets or confections. The guide, however, reached into his carrying bag, removed a jar containing grains and munched them leisurely. He took time to chew very thoroughly before swallowing. The others laughed at the guide, saying that grains could hardly give him the energy of sugary bonbons. But Evelyn I. marvelled at the Swiss guide's youthful vitality and energy. It was a contrast to the "quick tiredness" of the tourists. Where the Swiss guide's raw grain snack gave him steady and prolonged youthful energy, the tourists' sugary candy bars gave them speedy bursts of energy and a quick letdown while the Swiss guide was in the prime of vigor.

**Secret Given to Evelyn I.** The Swiss guide told Evelyn I. that many natives looked upon *raw grains* as a natural source of "instant energy" that revitalized the body and mind and produced an alertness that rivalled that of a youngster. Most importantly, the raw grains gave a steady supply of energy that did not have the up-and-down peaks such as produced by sugary confections. He told Evelyn I. that the Swiss consider raw grains to be "Nature's only sugar" with a powerhouse of youth-building energizers. He said it was traditional for the Swiss to use raw grains for snacks as well as for breakfast to help give them the stamina and youthful vigor needed to survive in the harsh climate. The secret was simple–substitute all "lifeless" packaged, processed and "enzyme dead" commercial cereals with raw grains. Add desired fruit, honey or milk, and enjoy "instant energy."

**Makes Evelyn I. Enjoy Energy of a Youngster.** Evelyn I. purchased a variety of different raw grains. She mixed them together. She munched them in place of candy, and it was a miracle how she soon felt "young all over" and enjoyed the energy of a youngster.

**How "Youth Enzyme X" in Raw Grains Produces "Instant Energy."** The energy from raw grain sugars becomes available to the body when the grain enzyme breaks glucose down in the tissues. The "Youth Enzyme X" in the grains promotes a form of natural oxidation that creates needed carbon dioxide and water in the system. The enzyme causes a *gradual* oxidation in a step-by-step level. The enzyme permits a small amount of energy to be released, using the step-by-step passage of the hydrogen from the carbohydrates to combine with oxygen to form a special group of compounds (high-energy phosphates) which are sources of youthful vitality.

The "Youth Enzyme X" in the raw grains then helps to break down these high-energy phosphates, so that their energy is released to the tissues. Here the tissues use the high-energy phosphates for the synthesis of other compounds that provide body heat and energy for muscular contraction or active physical movement. All of this is sparked by "Youth Enzyme X" found in raw grains. The energy materials from the grain itself are used by the enzyme in the grain. Truly, a miracle of Nature–in a small living grain!

### *How "Raw Grains" Help Improve Digestive Function*

Traditionally, the hardy, sturdy people of Europe look to raw grains as a natural healing source. Often they will eat a handful or two of raw grains together with a hot beverage. This helps create a more palatable taste. But the Swiss have long favored *chewing raw grains* very thoroughly. Mouth enzymes thus predigest the grains, and help extract valuable nutrients that are released from the starchy cellulose envelopes so they can promote speedy healing of digestive disorders. The seemingly ruggedly youthful people of Europe and Switzerland rely upon a traditional natural "raw grain digestive tonic" that is made as follows:

Make a mixture of two tablespoons of rye berry kernels, two tablespoons of raw wheat germ, one tablespoon of millet flakes (sold at most health stores or large food outlets), three shredded figs, and a handful of sun-dried raisins.

Place this entire mixture into a cup of water that has been boiled but is now comfortably lukewarm. (Do NOT let this Raw Grain Digestive Tonic boil because that would destroy the valuable "Youth Enzyme X" and other nutrients.) Mix thoroughly and then eat with a spoon.

*Digestive Benefit:* The miracle "Youth Enzyme X" takes out valuable protein from the raw grains, blends it with the grain "sugar" and combines it with fructose (natural fruit sugar) to send a stream of revitalizing nutrients into the digestive system.

The now-metabolized amino acids enter into the building and rebuilding of the digestive tract cells and tissues. This helps boost the youthful efficiency of the digestive system and is regarded by the Swiss as the key to better health.

Folk healers have taught the Swiss that the Raw Grain Digestive Tonic will give them "youth" in their "middles." What you eat is important, say the Swiss, but what you assimilate is youthful! As living proof, we look to the Swiss people. Men and women in their sixties, seventies and eighties have amazing vigor. They are hardy, vigorous, mentally and physically alert. They are mountain climbers, manual laborers, and woodsmen. The men and women do heavy work from sunrise to sunset. They have fewer of the modern conveniences than other Europeans, but more health and vigor than many other nationalities. The secret may be in their reliance upon raw grains with the magic "Youth Enzyme X" factor.

### *How the Swiss Use Grains for More Youthful Living*

Swiss folk remedies place importance upon using raw grains for a variety of youth-building benefits:

*Quick Energy:* Chew or munch any raw grains, singly or in combination. Thorough mastication is essential to release the valuable enzymes and energy-producing carbohydrates.

*Gastrointestinal Rebirth.* Raw corn kernels or powdered corn have a fine binding action that is smooth and gentle on the gastrointestinal organs.

**Restoration of Regularity.** Add raw barley grains to fresh vegetable juice, stir thoroughly and drink as a natural gruel. It is said to have a demulcent, soothing benefit on the intestines, to soothe nervous indigestion and promote regularity–without harsh chemical laxatives.

**Blood Building.** Rye kernels thoroughly chewed and washed down with fresh fruit juices send a stream of blood-building minerals into the system to help put color into your cheeks and warmth in your fingers and toes.

**Skin-Hair Health.** Millet in powdered flake form (sold at many supermarkets and health stores) is a powerhouse of silicon and other minerals that go into the regeneration of skin cells and hair follicles. Swiss folklore suggests using raw millet flakes regularly to have a youthful skin and hair growth.

**Circulation Booster.** Raw oats that are chewed thoroughly by themselves, or mixed into fruit juice or milk according to taste, are said to be a stimulation booster to the circulation. The "Youth Enzyme X" factor works up a sluggish blood circulation and helps increase sleepy processes that vitalize the body from head to toe.

**Respiratory Rejuvenation.** Raw buckwheat is a source of rutin, a nutrient that works together with vitamins to help strengthen body cells and tissues and improve the health of the organs of respiration. Buckwheat rutin helps strengthen the fragile cells of the nose-throat-chest membranes to resist allergies, colds, and infections. The Swiss remain amazingly free of winter ailments in the face of cold, snow and icy winter seasons. Folklore has it that chewing of raw buckwheat daily, or sprinkling this raw grain on salads, has the power to build resistance against winter ailments.

Raw grains form the foundation for youthful health for the seemingly "forever young" vigorous Swiss. They are slowly gaining recognition throughout the world. Raw grains are prime sources of the magical "Youth Enzyme X"–the natural way to rebuild youth and health from within.

*Special Points to Remember:*

1. The magic "Youth Enzyme X" is found only in raw grain foods that are nonprocessed and chemical-free.
2. For a source of super-powerful enzymes, try the Swiss secret of germinated seeds.

3. A variety of different raw grains, each with a unique taste thrill, makes "Youth Enzyme X" available in endless combinations.

4. The raw grain breakfast, Swiss Muesli, gave Gloria G. a supercharging of enzyme-powered vitality and energy.

5. Evelyn I. learned the secret of "instant energy" from a vigorous "over-sixty" mountain climber who chewed raw grains for endless vigor.

6. Swiss folk remedies, using raw grains, help remake the body from head to toe, thanks to the "Youth Enzyme X."

# 8

# *How Food Enzymes Offer Protection Against Colds, Allergies, and Respiratory Disorders*

Enzymes in raw foods as well as in the body are able to create a magic-like protection against colds, allergies and many respiratory disorders. The extreme sensitivity toward cold-catching that plagues many older persons could be eased and transformed into resistance and "bounce back" vigor of youth with an adequate intake of simple yet effective food enzymes. A program of wholesome, natural foods is a treasure of these allergy-fighting enzymes that help build a bulwark against the invasion of infectious bacteria.

## *How Enzymes Use Vitamin C as a Natural Cold Preventive*

Enzymes take Vitamin C from fresh raw foods and use it for the production and repair of a strong cement-like substance known as *collagen.* Enzymes enable Vitamin C to make collagen for use as a chief component of all connective tissue, tendons, cartilage, bones and skin. If there is an enzyme deficiency, Vitamin C loses its ability to make collagen. The body's supply of collagen weakens; the connective tissue starts to break down. The bones become brittle. The blood vessel walls of the respiratory tract (and elsewhere in the body) lose needed elasticity. Once this collagen breakdown occurs, the body becomes more sensitive to the invasion of external bacteria. Infectious germs enter the body, seize hold of the weakened and broken tissues, embed themselves and then start the toxic reaction that erupts in the form of a cold.

With a regular supply of raw food enzymes, the tissues are able to have strengthening collagen from Vitamin C and are able to build and rebuild their structures to resist infectious bacterial invasion. But without enzymes, Vitamin C remains inert, lifeless, unable to function. Nature has put enzymes and Vitamin C in fresh fruits and vegetables in a delicate but potent balance that has been seen to provide youthful resistance and recovery powers from colds and other respiratory disorders.

### *How the "Right Fruit" Gave a Clerk a Magic Freedom from Colds*

Joanne McK. worked in an overheated office for most of the day. No doubt the close confinement and poor ventilation did much to reduce her resistance to cold-catching as well as the troublesome sniffles throughout the winter. But her problem was that she followed the wrong program in building her body's natural supply of enzymes and Vitamin C for fighting colds. She knew that fruits and vegetables were excellent natural sources of these substances. So Joanne McK. ate plenty of them throughout the year. But she ate *cooked and processed* fruits and vegetables. They were devoid of perishable enzymes and Vitamin C. She kept catching one cold after another from early autumn to just before spring. She often lost days of work because of the weakening effect. That was when she made a minor but effective change to the "right fruit."

**Simple Enzyme-Vitamin Fruit Program for Cold Resistance.** Each morning, Joanne would eat a *fresh raw fruit salad* and also have a glass of a *fresh fruit juice.* At lunchtime she would eat fresh *raw* fruit and also have a glass of a fresh *raw* juice. At dinner she would have a raw fruit salad with her main meal and have another fresh *raw* juice. It was this simple enzyme-vitamin program that fortified her body to resist chronic and habitual cold-catching. The benefit was seen in her ability to enjoy freedom from colds. But more important, if she did experience some sniffles, she just boosted her intake of the "right fruit" that was *raw, uncooked, and nonprocessed,* and this helped her bounce back with speedy recovery.

**How Enzyme Vitamins Build Natural Protection.** Many conditions of colds, allergies and related respiratory ailments are traced to the body's deficiency of antibodies. (These are disease-fighting mechanisms formed by enzymes in the presence of infection and are sent through the bloodstream to neutralize the infectious bacteria causing colds.) Enzymes are needed to use Vitamin C to stimulate the white blood cells into using more enzyme-created antibodies to destroy harmful organisms that lead to colds and respiratory infections.

Enzyme vitamins work to form new cells, form a steady supply of natural protective antibodies and to replace impaired tissues. In many fresh raw fruits and vegetables, the cold-fighting enzyme known as ascorbic acid oxidase (or ascorbase) exists together with Vitamin C to create a magic cold-fighting function. Ascorbase changes Vitamin C into dehydroascorbic acid which is a potent source of healing and neutralizing infectious bacteria. Nature has placed enzyme vitamins in raw fruits and vegetables to help build natural protection against bacterial invasion. In the raw fruit or vegetable or in the freshly prepared raw juices, enzyme vitamins are Nature's best and safest method of all-natural immunity.

### *How Enzyme Vitamins Promoted Freedom from Allergies*

Years of allergy attacks left George N. looking and feeling older than his middle forties. As a technical researcher, he was constantly leafing through shelves of books and files with the occupational hazard of dust and mold. The slightest spray of dust from an old book would send him into a sneezing and sputtering spree. George's eyes watered. He could scarcely breathe. He thought he was choking. He would stagger toward a window, open it wide, and breathe deeply. His problem was intensified because he had to gulp ice cold air in winter, not to mention thick swirls of pollution. His respiratory tract became clogged and malfunctioning so that his allergies worsened. He had asthma-like distress, bronchial spasms and painful breathing. George N. was pale, shaken and prematurely aged because of this allergy-prone problem.

**Drug Reaction Leads to Natural Protective Measures.** He reacted adversely to drugs so was forced to seek other treatments. A friendly pharmacist told him that many patent remedies were effective because of the high supply of Vitamin C in the formula. It was so stated on the list of ingredients. The pharmacist suggested that George N. protect himself against allergic sensitivities with a high enzyme vitamin intake. The program was amazingly simple and just as amazingly effective. It called for eating lots and lots of fresh *raw* fruits and *raw* vegetables throughout the day; it also called for drinking lots of the *raw* juices from the same source: *raw* fruits and vegetables.

**The Natural Antihistamine Tonic.** Mix together equal portions of orange, grapefruit and tangerine juice. Add one tablespoon of rose hip powder. (The fruit or seed pods of the rose flower are known as rose hips. It is available in powdered form at many health stores and large pharmacies. It is a natural form of one of the richest known sources of enzyme vitamins; it is a treasure of allergy-fighting bioflavonoids and Vitamin E, as well as blood-building minerals.) Stir together and drink one glass in the morning, another at noon, a third glass in the afternoon and a fourth glass in the evening.

*The Magic Power of the Tonic:* The enzyme vitamins in this tonic created a natural antihistamine action. During an allergy attack, the body seeks to reject the offending and foreign bacteria that have invaded the bloodstream. The body sends out a supply of *histamine* through the tissues. A problem here is that an excess of histamine causes an imbalance and this is the cause of much of the allergic sputtering and choking. An antihistamine ingredient will neutralize the excess amount and help relieve symptoms. The enzyme vitamins in this tonic are able to create a *natural antihistamine reaction* by opening up the air passages in the lungs, stabilizing stomach processes, and normalizing the amount of blood to the damaged tissues. Thus a natural healing process takes place.

George N. took this Natural Antihistamine Tonic throughout the day. Together with the enzyme vitamins in the fresh raw fruits and vegetables, he was able to build his own supply of antibodies that worked to inactivate infectious microbes and their age-causing toxins.

He soon recuperated so that dust and molds had little allergic effect upon his breathing. Once his freedom from allergies was enjoyed, this technical researcher looked young, had a clear skin with bright color, had bright eyes, enjoyed vigorous mental and physical alertness. He enjoys life once more, thanks to the regular and all-natural intake of enzyme vitamins and his regular *Natural Antihistamine Tonic.*

### *How Enzyme Vitamins Build Youthful Immunity to Respiratory Distress*

Enzyme vitamins found in raw fresh fruits and vegetables are especially beneficial to older folks who need youthful immunity to respiratory distress. These enzyme vitamins are very highly concentrated in the adrenal cortex which reacts to conditions of stress as well as normal aging. Reportedly, the enzyme vitamins are rapidly depleted in the glands during middle age when the body weakens and undergoes many stress or tension-trauma situations. This renders the person more sensitive to respiratory ailments in the decline of much-needed enzyme vitamins such as Vitamin C. Many such older persons have low plasma levels of enzyme vitamins. Taking enzyme vitamins from fresh raw fruits and vegetables will offer these youthful immunity-building benefits:

1. Enzyme vitamins strengthen cells and tissues to prevent the predisposition to gastrointestinal bleeding and loss of protective allergy-fighting antibodies.

2. Enzyme vitamins increase resistance to acute hemorrhage and are therapeutically effective in traumatic shock. Both situations deplete supplies of allergy-fighting substances and need to be corrected.

3. Enzyme vitamins reduce the permeability of the capillary vessels, thus reducing the normal inflammatory process. *Benefit-Plus:* Because enzyme vitamins reduce the permeability of cell membranes, they help inhibit cell penetration by infectious bacteria and virus germs. This is the unique and powerful allergy-fighting benefit of enzyme vitamins.

4. Enzyme vitamins act as spark plugs to leucocytes (white blood cells) to surround and dispose of invading viruses and bacteria. These enzyme vitamins actually supercharge leucocytes to promote this internal cleansing action to build immunity and create freedom from allergic attacks.

5. Enzyme vitamins work to build hemoglobin (the red coloring of blood). It is hemoglobin that carries oxygen to all the cells in the body, with the iron necessary to build and rebuild tissues to resist infectious invasions.

*Note:* Your body must have *living enzyme vitamins* if it is to benefit from these youthful immunity-building actions. Nature has put these living enzyme vitamins in fresh raw fruits and vegetables for your own benefit. Use them to help build immunity to allergic disorders.

### *Good Sources of Enzyme Vitamins*

Oranges, grapefruits, tangerines, strawberries, cantaloupes, broccoli, Brussels sprouts, dark green leafy vegetables, cauliflower, tomatoes, raw sweet peppers, raw cabbage, sauerkraut, and potatoes are good sources.

The pulp of orange and grapefruit is especially high in enzyme vitamins. Eat the whole fruit often.

*Special Tip:* The sweeter the fruit (whether pulp or fresh raw juice without sugar), the more potent it is in enzyme vitamins.

### *How to Protect Delicate Enzyme Vitamins in Foods*

When a fruit or vegetable is peeled or chopped, the perishable enzymes slowly combine with oxygen and lose their potency. Since enzyme vitamins dissolve in water, you run the risk of losing much potency when you wash, soak or boil such foods. To help protect delicate enzyme vitamins, here are some suggestions:

1. *Wash Quickly.* If you must wash fresh raw fruits or vegetables, do so quickly. Avoid soaking. Plan to buy and use these foods as quickly as possible with a minimum of preparation.

2. *Refrigerate.* Always refrigerate fresh fruits and vegetables before use. Place in a covered container or plastic bag. Freezing inactivates the enzyme vitamins but will not deplete them.

3. *Greens.* The darker the color of vegetable greens, the more potent the supply of enzyme vitamins. But cutting and exposure to air will deplete potency. So plan to cut or shred greens just before use. If possible, eat whole greens with a minimum of cutting: this helps exercise your jaws and improves the flow of salivary enzymes that work with vegetable enzymes to metabolize valuable nutrients.

4. *Cooking.* Fruits should be eaten raw for their maximum enzyme vitamin supply. Vegetables should also be eaten raw, but those that must be cooked should be put in a small amount of boiling water, cooked covered until just tender. *Tip:* Use the vegetable juice as a healthful broth.

5. *Potatoes.* These are a good source of enzyme vitamins. Cook potatoes in their scrubbed skins and then eat the skins too.

6. *Cutting.* Avoid using any knife containing copper since this metal destroys enzyme vitamins.

**Boost Enzyme Vitamin Intake with Fully Ripened Fruit.** Nature increases the enzyme vitamin content of fully ripened fruit. Underripe fruit has a weak amount of these needed cold-fighting enzyme vitamins. If you purchase underripe fruit, let them remain at room temperature in a dark place until well ripened. (The exceptions are fruits with thick peel such as those in the citrus fruit group, bananas, or apples: they may be put in a refrigerator for ripening. Their thick peel protects against oxygen depletion of enzyme vitamins.)

### *How Fermented Milk Enzymes Offer Natural Cold Protection*

Eileen Q. has been able to reduce the constant wheezing and coughing of respiratory distress. She has discovered that fermented milk (yogurt, buttermilk, sour milk, sour cream, cottage cheese) is soothing to her respiratory tract. She has fewer symptomatic reactions from allergic attacks when she fortifies her body with enzyme vitamins and substances found in fermented milk. She feels more energetic and alert. She can breathe better too.

**Secret Enzyme Benefit of Fermented Milk.** Respiratory ailments often reduce the interior spaces of the breathing apparatus, especially the nostrils. This constriction prevents oxygen from circulating. Now the cold-causing organisms and allergic bacterial wastes are able to take hold and breed and cause discomfort. But enzyme minerals in fermented milk combine with enzyme vitamins to strengthen the nasal cilia (tiny microscopic hair-like things which cover the mucous membrane and filter out offensive allergy-causing foreign substances.) The cilia must be moist, strong and ever-resilient. This is made possible by fermented milk enzymes, vitamins and minerals that prompt the various mucous membranes to issue forth a substance called *lysozyme,* which is Nature's miracle allergy-fighter.

Without the presence of sufficient enzyme vitamins and minerals, *lysozyme* production falls off and the body renders itself susceptible to allergic disorders and cold-catching. A deficiency of fermented milk enzymes may cause dryness of the cilia cells, inhibiting the *lysozyme* production and lowering the natural antiseptic secretions that normally protect against infections. The key to the strength and health of the cells of the mucous membranes is in fermented milk enzymes. They help to rejuvenate and replenish these cells when they have been injured or destroyed by attacks of germs.

**Fermented Enzymes Better Than Plain Milk.** Fermented enzymes found in yogurt, buttermilk, sour milk, sour cream, or cottage cheese are powerhouses of *lysozyme*-producing agents. They work better than plain milk for the simple reason that bottled milk is precooked, boiled, homogenized, processed and subjected to storage and shipping. Plain milk has no living enzymes. They have been destroyed by heating and processing. The vitamins and minerals in plain milk may be sluggish or weak since they need enzymes for function. Without enzymes at all, milk vitamins and minerals would be undigested and lie in a mass in the system. A weak body enzyme network means weaker digestion of milk vitamins and minerals. But *fermented milk* is a source of natural enzymes which are alive and active: they take up the vitamins and minerals from the fermented milk and help build protective substances in the body that add up to youthful health and resistance to colds and allergies.

For many, a simple substitution in dairy products can do much to fortify the body with living enzymes and living vitamins and minerals in the form of fermented milk products, available at just about every dairy food outlet and health store in the country.

### *A Folklore Enzyme Healer for Allergies*

Klara Y., a European houseworker, amazed the household by rarely catching cold and resisting contagious diseases when others had serious bronchical disorders and winter ailments.

Klara Y. said that it was a folklore remedy to chew and eat garlic daily. It was said to help build resistance to infections and act as a natural healer for respiratory or bronchial disorders.

Other members of the household began taking this folklore healer and were able to heal some of their allergic symptoms, but because they disregarded the other natural laws of healthful living which included fresh fruits and vegetables and natural foods, their benefits were proportionately reduced.

**Enzyme Healing Power of Folklore Garlic.** Raw garlic contains natural plant enzymes which propel the plant's supply of oil into the digestive-assimilative system. Enzymes then unite the garlic oil with the body's own supply of sulfides and disulfides (allergy-fighting compounds) and send this mixture to the virus matter. The harmful effects of the virus organisms are inactivated and a healing power is experienced. For many Europeans, eating garlic is a natural way to help fight allergic disorders ranging from the common cold to bronchial spasms and attacks.

Garlic enzymes thereby help to neutralize viral infections and cleanse the system so that it can enjoy youthful health and vigor. It is an Old World folk healer that is fast gaining acceptance in the New World in the quest for the natural way to prolong youth and heal illness.

### *A Doctor's Program for Curing the Common Cold*

A physician reports that he has found the all-natural and drugless way to cure the common cold. Documenting his findings in his writings,[1] Edmé Regnier, M.D. outlines a home program for treating colds. He suggests the taking of Vitamin C in supplement

1. Edmé Regnier, M.D., *There Is A Cure for the Common Cold,* West Nyack, New York: Parker Publishing Co., Inc., 1971.

form immediately upon recognizing the symptoms of a developing cold. Dr. Regnier's program calls for ten to eleven days of home treatment and is divided into four stages. It is simple and effective, according to this practicing physician.

*Stage One:* At the first sign of a cold, take 750 milligrams of Vitamin C; thereafter, take 625 milligrams at three-hour intervals until bedtime; take 750 milligrams before retiring; take 750 milligrams on arising; resume taking 625 milligrams every three hours throughout the day. Follow this program for three to four days.

*Stage Two:* Maintain the same timetable of home treatment, but reduce daytime doses of Vitamin C supplement to 400 milligrams; reduce bedtime and getting-up doses to 500 milligrams. Follow this program for three days.

*Stage Three:* Maintain the same timetable of home treatment, but cut daytime doses to 200 to 250 milligrams of Vitamin C. Cut bedtime and getting-up doses to 300 milligrams. Follow this home treatment for three days.

*Stage Four:* For one day, take 250 milligrams of Vitamin C every four hours during waking hours; on the last day, take 250 milligrams of Vitamin C every six hours during waking hours.

*Benefits:* Dr. Regnier has found that the common cold as well as many respiratory ailments could be eased and suppressed by the intake of Vitamin C. It is believed that enzymes will be able to create healing with larger amounts of readily available Vitamin C; that is, in addition to food intake sources. Dr. Regnier cites many cases of miracle recovery from colds and permanent cure through the use of enzyme-propelling Vitamin C. The program is natural. According to Dr. Regnier, it is curative. Supplementation may well be the double-barrel effect sought for rejuvenating the body so it can resist-recuperate from respiratory infections.

### *How Vegetable Enzymes Help Rejuvenate the Respiratory System*

Enzymes in raw vegetables are known for having valuable disease-fighting powers. These enzymes take out Vitamin A from the raw vegetables and use this nutrient to improve the antibody-making powers of the bloodstream. The key to rejuvenation-healing of the respiratory system is in the ready availability of infection-fighting antibodies. By eating lots of fresh raw vegetables that contain Nature's built-in enzymes, the system is doubly rewarded by having these magic substances feed Vitamin A to the deficient bloodstream.

Vegetable enzymes use Vitamin A to nourish the mucous membranous tissues that line the body openings such as the nose, throat, sinuses, middle ears, and lungs as well as the organs of elimination.

These raw vegetable enzymes use Vitamin A to energize the membranes so that they can secrete the liquid covering for the cells that prevents bacteria from reaching within. Furthermore, bacteria cannot live in mucus. Enzymes offer this protective function of keeping the cells moist to create an unfavorable habitat for bacteria. Enzymes use Vitamin A to digest worn-out tissues and remove the wastes. Enzymes see to it that potential disease-causing bacteria cannot reach the cells because of the protective moisture covering; or if they do take hold of the cells, enzymes create more moisture so that these bacteria are made ineffective.

Enzymes also work with Vitamin A from raw vegetables to cleanse the air sacs of the bronchial tubes and lungs. This "washing" helps promote better breathing and freedom from recurring respiratory infections.

A deficiency of raw vegetable enzymes and Vitamin A may cause an accumulation of much debris so that the mouth is dry, the membranes lose their protective moisture coating and the nasal channels become vulnerable to an invasion of bacteria. To help resist such a possibility, raw vegetables should be eaten daily.

**Food Sources of Vegetable Enzymes.** Carrots, beet greens, collards, dandelion greens, endive, kale, lettuce, mustard greens, parsley, green peppers, winter squash, sweet potatoes, tomatoes, turnip greens, and watercress contain valuable enzymes. *Note:* Vegetable enzymes and Vitamin A are more potent and vigorous in the spring and summer, when these foods are more readily available. Winter storage, together with the lack of sunshine and warmth, means the plant may be deficient in many nutrients. So if you want to get the most from raw vegetable enzymes and Vitamin A, eat these foods abundantly during the spring and summer. Eat copiously throughout the winter too. Use raw vegetable salads almost daily. Steam vegetables slightly, briefly, in a bit of water and be sure to use the liquid as a rich source of enzymes and Vitamin A.

Food enzymes are the natural way to enjoy a magic-like protection against colds and respiratory ailments. You will look and feel young if you can breathe healthfully. Look to enzymes as the natural way to enjoy freedom from colds and better youth and health.

### *Essential Points:*

1. Enzymes use Vitamin C in raw fruits to offer built-in protection against colds. Joanne McK. ate the "right fruit" for welcome healing and resistance to colds.

2. A simple enzyme program helped George N. enjoy renewed youthful freedom from allergic attacks.

3. A *Natural Antihistamine Tonic* helped this researcher's body build its own supply of infection-fighting antibodies.

4. Enzyme vitamins have five benefits that offer youthful immunity to respiratory distress.

5. Fermented milk enzymes gave Eileen Q. youthful protection against colds.

6. Klara Y. used a garlic enzyme to heal and resist allergies.

7. A doctor offers an all-natural program for curing the common cold.

8. Raw vegetable enzymes help rejuvenate the respiratory system.

# 9

# *Herbs: Nature's Garden of Youth-Building Enzymes*

The history of herbal rejuvenation is as old as man himself. Our current awareness of the healing power of Nature's grasses is a product of many thousands of years of observation and experimentation of people all over the world. Modern medicine is an outgrowth of man's efforts to help use natural means for healing of ailments. Long before the advent of the laboratory or the hospital, man sought help in the form of green medicine from herbal plants that grew all around him. Man slowly learned that herbs could soothe, heal, refresh and rejuvenate the body from head to toe. He began to identify these herbs and learned which ones would act upon specific parts and functions of the body and mind. As the centuries passed, knowledge of herbal rejuvenation passed into a science of natural healing. It formed the basis and foundation for modern medicine.

## *From Superstition to Modern Healing*

Ancient man recognized that healing as well as rejuvenation could be possible by using Nature's grasses for treating the *cause* of the problem. His medicine man told him that if he wanted to keep well and to remain youthful, he should look to the fields for secrets of natural rejuvenation. The ancient medicine man may not have known about vitamins, minerals or enzymes as we know of them today, but he knew that herbs promoted miracles of rejuvenation and he prescribed their use for daily living.

The ancient medicine man used religious dances and incantation uals (forms of spiritual prayer) as part of the ceremony of the e of herbal treatments. It was a form of helping the mind adjust to the possibility of physical healing. It also helped keep the mind youthfully alert while the body was undergoing rejuvenation through prescribed herbal treatments.

The superstition gradually evolved into a healing science as man recorded observations of animals and their healing programs. From markings made on stone, wood and papyrus, we learn that man watched ailing animals who would chew specific plants as healing medicines. Many sick cats or dogs purged themselves with certain herbal grasses. Still many other animals would roll in certain types of foliage to heal snake or insect bites. One repeated inscription was that most foods (grasses, plants, flowers, leaves, fruits, vegetables, seeds, nuts) which animals ate were said to be good for man. Slowly the different types of plants were catalogued. Their youth restorative functions were specifically recorded. Out of this beginning, modern medicine was born.

### *Healing Herbs of Science*

From a medicine man's superstition, herbs have come to the laboratory of science. For example, many modern "discoveries" or "breakthroughs" are based upon the use of herbs for youthful healing. To soothe the alimentary and bronchial tracts, onions and garlic are used as natural antiseptics. To soothe liver disorders, the dandelion is a natural medicine. To detoxify the bloodstream, wild sarsaparilla is known for having cleansing enzymes. The digestive-intestinal organs are cleansed by masterwort. Many medicines that heal ulcers or skin lacerations have cinquefoil stems as a base. To treat blood congestion, unsaturated safflower oil is often part of the prescribed medicine.

### *Herbal Enzymes That Help Restore Youthful Health*

Each people has its history of herbal rejuvenators. The Orientals prize *ginseng root* for its store of youth-building enzymes. The Anglo-Saxon peoples look to *chamomile* as an herb with enzymes that help folks bounce back with youthful vigor from colds and stomach disorders. The natives of the West Indies have faith in *mauby bark* for rejuvenating and replenishing the bloodstream and also using *ginger root* to boost resistance against respiratory distress. Many Hungarians use a folklore gypsy herbal healer for rejuvenation; the gypsies prescribed the leaves of *sweet chestnut* for healing debilitating asthma, *parsley* to soothe a disturbed bladder, *stinging nettle* to regulate blood pressure and "softening" hardened arteries, *rosemary* to boost a youthful hair growth, *milkweed* for detoxifying the bloodstream. The Irish place faith in the use of *leek* (pounded to a pulp) as a natural poultice for inflammed eyes; they use boiled *laurels* and *ivy leaves* to heal burned skin; to heal colds, they boil *seaweed* with *wildflowers* and use as a tonic.

The healing lore of all peoples from almost all parts of the world has become the foundation of modern medicine. Modern medical doctors will often recommend tested herbal remedies for patients who show drug reactions or who seek a natural way to better health. Herbal healings may have originated with witchcraft or devil worship, but they are known to work and are gaining acceptance in modern science. Many plant derivatives of herbal plants are included in such official compendia as the *United States Pharmacopoeia* and the *National Formulary.* Modern medicine owes its success to many so-called "superstitious" plant healers of the ancient witch doctor. Some examples are:

| *Herbal Medicine* | *Healing Functions* |
|---|---|
| Burdock | Increases perspiration and helps wash out wastes from the body. |
| Cascara Sagrada | Acts as a natural laxative. |
| Cinchona | A source of quinine and used to heal malaria. |
| Dandelion | Contains enzymes that promote regularity and also help detoxify the liver. |
| Digitalis | Acts as a diuretic and also helps stimulate the heart in cardiac unrest. |
| Ephedra | A source of ephedrine which is soothing for allergic disorders including hay fever, asthma, bronchitis and sluggish circulation. |

| | |
|---|---|
| Gentian | Helps in problems of digestive disorder and dyspepsia. |
| Goldenseal | Helps soothe problems of gastrointestinal upset and also helps in respiratory congestion. |
| Juniper | Used to stimulate sluggish processes and also a natural diuretic. |
| Licorice | Soothes irritated mucous membranes and problems of excess acidity. |
| Lobelia | A natural healer for respiratory problems such as asthma or bronchitis. |
| Mandrake | Helps to cleanse the liver and gall bladder. |
| Sarsaparilla | Helps cleanse toxic wastes in the bloodstream and also as a poultice to reduce inflammatory swellings on the body. |
| Thyme | Enzymes exert a therapeutic healing on the bronchial tubes. It is often part of a prescription product that helps sooth spasmodic, irritating coughs and winter ailments. |
| Witch Hazel | Has astringent-antiseptic qualities to soothe feverish reactions. Has quick hemostatic action (checks bleeding) and promotes speedy healing. |

**Superstition or "Faith" Helped Boost Herbal Healing.** The tribal healer, whether a medicine man, shaman, voodoo priest or witch doctor, was wise in knowing that faith was an essential part of healing. He used songs, incantations, mysterious spells, secret ceremonies and magic as a means of creating an attitude that the divine spirits were gathering around to help promote healing. Once he instilled confidence in the ailing person, the magic witch doctor then administered the healing herbs. In such a mystical environment, a climate of "faith" was created. It was a primitive form of psychosomatic healing, accompanied by herbs that was largely responsible for the recovery of the patient.

The witch doctor did not praise himself for healing. He told his patients that all knowledge and healing came from divine spirits and the knowledge of magical herbs. He explained that his magic incantations and mystical spells helped drive away the sickness-causing evil spirits. He pointed out that the sacred fires that often burned during such white magic ceremonies helped cleanse the air of devils and permitted good spirits to enter the compound and thereby enter into the body of the afflicted. He wrapped his white magic ceremonies in mystical knowledge and lore. He was successful in many reported ceremonies. He knew that herbs possessed magical therapeutic values and they were used to give new life to his followers. The witch doctor was the forerunner of modern medicine that has rediscovered the magical healing powers of herbs of the fields. Together with our modern knowledge of psychosomatic faith and confidence in the physician, an ailing person has hope for recovery and rejuvenation.

## *How to Use Herbs for Enzymatic Rejuvenation*

*How to Buy Herbs:* Ask your local health food store for your desired herbs. Inquire at your local herbal pharmacy. Look in the classified telephone directory under "herbs" for nearest available suppliers. Ask local organic farmers if they have herbs available for sale; if not, ask them to recommend a source. Ask your local Department of Agriculture representative for sources of herbs. For maximum enzyme effectiveness, herbs should be organic, non-processed, and nonadulterated. Enzymes are more potent in *raw* herbs.

*How to Use Herbs:* Just one pinch will suffice. To release the effective enzymes, crush the herbs in the palm of your hand with your fingers or a wooden spoon. Then sprinkle over a *raw* fruit or *raw* vegetable salad. Sprinkle over raw grain cereals. Mix a pinch of an herb in a fresh *raw* fruit or vegetable drink and sip slowly. Because enzymes are destroyed when the herbs are cooked, it is advisable to use them for uncooked meals.

*How to Store Herbs:* Keep in a glass jar, sealed, in a cool, dry, dark place. Avoid placing in any metal container as the acids from the metal may corrode the herbs and destroy their enzymatic power. As soon as you have removed herbs for use, quickly seal the jar and return it to its cool, dark place. While *fresh* herbs have maximum enzyme power, you may store them up to several months and still have vigorous healing benefits from these grasses.

### The Healing Secrets of Herbal Enzymes

Herbs are Nature's plants that grow in rich soil. The roots of these plants absorb the nutrients in the soil as provided by the air, sun and water. In addition to vitamins, minerals and plant proteins as well as essential fatty acids, herbs contain a rich treasure of enzymes. As raw foods, they are prime sources of these valuable youth-building enzymes. Rediscovered secrets of their rejuvenation actions have made herbs a source of natural and magical healing. Herbs reportedly offer these healing-rejuvenation secrets:

*Volatile Oils:* Boost enzymes in the mouth and stomach. These oils use herb enzymes to activate peristalsis and reduce discomforts of fermentation; they also exert an antispasmodic and anodyne (pain-relieving) benefit. These herbal oil enzymes boost digestion and act as gastric tonics. They also help to dilate cramped blood vessels (vasodilation) to help promote a more youthful blood-stream.

*Tannins:* These are enzymatic agents that have an astringent and antiinflammatory benefit on the gastrointestinal system. They promote a soothing relaxation of these organs and facilitate better healing.

*Tonic Principles:* The pungent taste of many herbal plants is traced to somewhat bitter-tasting enzymes that exert a tonic effect on the smooth muscles of the gastrointestinal networks. These enzymes help perk up a sluggish digestive system and promote better appetite as well as better assimilation of foods that are eaten.

*Secretins:* These are enzyme substances that enhance the hormonal flow of many body glands, specifically the pancreas. Enzymes promote a youthful hormonal flow to give the feeling of rejuvenation from the inside to the outside. Enzymes promote this secretion reaction to sluggish body glands. They are natural hormonal activators.

*Organic Acids:* These are enzymes found in herbs that help stimulate the appetite, stabilize digestive functions, and appear to enter into the metabolization of fats.

*Alterative:* Herbal enzymes help to bring about a natural healing action so that minor functional disorders may be corrected.

*Antispasmodic Action:* Herbal enzymes (which form the basis for pharmacological medicines) are known for helping to allay or build natural resistance to the recurrence of spasms. These enzymes alert valuable vitamins and minerals to offer nourishment to parched muscular-cellular tissues to help ease problems of neuritis, irritability and excessive contraction. Many have hailed herbal enzymes as all-natural tranquilizers.

*Expectorant:* Plant enzymes promote a corrective homeostasis in the system to further help the body expel mucus from the respiratory tract. It is the natural way to promote autolysis of accumulated mucus that inhibits free aeration of the cells and tissues.

*Diaphoretic:* Herbal enzymes send nutrients into the respiratory system to boost a sluggish circulation; this helps increase natural perspiration and thereby help eliminate waste substances and toxic accumulations through the skin pores.

*Nervine:* Herbal enzymes use minerals to bathe the arterio-neural components of the body and thereby help to quiet a temporary nervous irritation caused by tension, fatigue, eyestrain, or headache. These enzymes create a soothing and quieting influence upon the nervous-emotional systems. Herbal enzymes are nonaddictive. They are not narcotic. The ancients recognized this benefit and freely prescribed herbal plants for natural healing of the nervous system.

### *Plant Enzymes: Roots of Youthful Health*

The ancients revered and worshipped the holy herbs for miracles of healing. Mysticism and devil worship often accompanied the prescribing and administering of herbal healers in sacred rites. It was this ritual that helped instill confidence in the patient. He believed that the witch doctor did communicate with the good spirits and with the use of these sacred youth herbs could give him healing and regeneration. It was a primitive form of self-confidence and herbal healing but it enabled civilizations to survive during an era when chemotherapy was unknown. Today, plants are gaining wide recognition in modern medicine. Herbology is also known as materia medica, botanicals, pharmaceuticals and even vegetable drugs. By any name, herbal plant enzymes are the roots of youthful health. Here is a collection of these youth-restoring herbs and their specific enzyme benefits.

**Acacia.** The gum of the acacia tree bark acts to remove phlegm from the throat and bronchial tubes. Enzymes ease inflammation and soothe the irritated mucous tissues of the throat.

**Agrimony.** The entire plant contains enzymes that help cleanse the liver of toxic accumulations and build a better blood circulation.

**Allspice.** The fruit, leaves and twigs have enzymes that help cool fevers and bring about speedier recovery from colds and winter ailments.

**Aloe.** The dried juice of the leaves helps correct chapped skin and promote better healing of skin infections such as insect bites.

**Angelica.** Enzymes in the leaves, seeds and roots act as a stimulation to the system to promote better perspiration. These enzymes help correct disorders caused by colds and winter ailments. These plant enzymes are said to soothe an irritated stomach and ease problems of colic.

**Balm of Gilead.** The buds are regarded treasures of enzymes that act as a natural expectorant in throat and breathing disorders. Many cough medicines contain this herb.

**Barberry.** The root bark is the most potent source of enzymes that cleanse the bloodstream and help soothe the gall bladder and the liver.

**Blackberry.** The leaves, fruits and roots are prime sources of enzymes that help to dissolve accumulated tartarous substances. The enzymes promote the action of citric and malic acids that help cleanse tartar accumulations on the teeth and in the mouth as well as upon the various body organs. It is a natural way to self-cleanse the body of impurities and encrustations.

**Burdock.** The leaves, fruits and roots contain enzymes that reportedly help detoxify the bloodstream. Water accumulations (known as edema, dropsy or simply water-logging) reportedly respond to enzymes in this plant. These enzymes help wash out water deposits from the tissues and cells.

**Cassia.** Herbal enzymes of the dried fruit act swiftly to establish regularity. This herb is often used in commercial laxatives.

**Catnip.** You can inhale the fragrant enzymes in the mintlike aroma of the leaves and flowering tops. These enzymes help soothe and relax the digestive system, refresh the circulatory system and create an overall feeling of well-being. It was often prescribed to help relax and then bring about a natural desire for sleep.

**Centaury.** The dried flowers are rose-colored and tasty. The enzymes act upon the kidneys and liver and help to cleanse body organs while washing out the bloodstream.

**Chamomile.** The upper half of the plant contains anticolic and gas-dissolving properties. Those troubled with gastrointestinal disorders due to gas bubbles trapped in the digestive system often find that chamomile helps dissolve these bubbles and promote better assimilation of food.

**Cherry (wild).** The trunk bark has lenticels (pores or tiny indentations) that are a prime source of enzymes that offer an astringent and healing property for external cuts and bruises. In particular, wild cherry enzymes are especially soothing for bronchial disorders because they help remove hardened accumulations of mucus which are responsible for coughs.

**Chive.** The hollow stems contain enzymes which offer a self-protective action against the corrosive effects of toxic wastes that infest the bronchial and digestive channels. Enzymes work upon chive to boost its blood-building sulfur and iron content. These enzymes then help stimulate the salivary and digestive enzymes to promote better assimilation of nutrients. *Folk Remedy:* A mixture of chopped chives, garlic and onion is said to be a natural cold-cough remedy. The secret here is that the chives' sulfur enzymes extract the garlic's active bactericidal crotonaldehyde to mix with enzymes of the onion and promote a natural antiseptic action against infections. It is said that this three-way enzyme action is a "miracle" cold fighter.

**Clover (red).** The rose-red flowering tops contain enzymes that reportedly help purify the bloodstream. These enzymes also help improve the production of necessary red and white blood cells to fight age-causing illnesses.

**Comfrey.** The glutinous leaves and roots have enzymes that act as an emollient and expectorant in conditions of bronchial distress. Many cough medicines contain comfrey. Enzymes help slough off catarrh from the tubes of the bronchial network. They help promote better breathing.

**Digitalis.** Also known as foxglove, this herb contains enzymes which help bring about youthful regeneration of the heart apparatus. Digitalis enzymes have been used in pharmaceuticals to heal disorders of the heart.

**Elm.** Chew the dried inner bark of slippery elm to release the bland mucilaginous enzyme that acts as an emollient and soothing demulcent to the system. It is said to ease coughing spasms.

**Fennel.** Enzymes abound in the sweet anise-like seeds and yellow flowers. These enzymes reportedly act as a gentle stimulant, dissolve gas bubbles naturally, soothe colic pains, and help relax nervous stomach complaints.

**Fenugreek.** The seeds yield mucilaginous enzymes that help soothe inflamed digestive organs, help correct intestinal unrest and may even be used as a poultice for wounds and inflamed regions.

**Figs.** A succulent fruit plant that was prized by the ancient Greeks and Romans for promoting better health and vigor. Enzymes in raw figs are said to have a soothing effect upon the digestive system. Many laxatives contain extracts of figs (enzymes).

**Flaxseed.** The gummy seeds are said to contain enzymes that help remove toxic waste accumulations from the bronchial canals and also to promote a cleansing action upon the kidneys. Helps create cleaner body organs and a better bloodstream.

**Garlic.** The leaves and bulbs (cloves) contain enzymes that help prevent infection and also act as an antiseptic on the intestinal canal. Digestion is boosted when garlic enzymes create a gentle but effective activation.

**Ginger (wild).** The roots are prime sources of enzymes which act upon the kidneys to help wash away viscid accumulations and help promote better hormonal function.

**Ginseng.** Orientals prize the enzymes in the root of this herbal plant for acting as a near cure-all. It reportedly is a powerful restorative, said to extend the prime of life and act as a natural aphrodisiac.

**Goldenseal.** American Indians worshipped the roots of this herb. It contains enzymes which reportedly heal the stomach, have a natural cleansing action to the colon, and stabilize the various body systems. The enzymes are said to promote better digestive ability.

**Hawthorn.** The dried fruits of this herbal plant reportedly contain enzymes that act as a natural diuretic; they are able to help wash out accumulated crusts. Specifically, hawthorn enzymes alert its crataegin and amygdalin to act as a natural heart stimulant and ease blood pressure unrest. Many modern chemotherapies contain extracts of this herb.

**Irish Moss.** The dried, bleached herb contains enzymes that act as a reportedly gentle emollient to soothe coughing spasms and help correct bronchial disorders.

**Juniper.** The ripe fruits are prime sources of enzymes which reportedly help wash out excess body water and also help to cast off unnecessary toxic deposits in the urinary system.

**Kelp.** Enzymes from this seaweed are beneficial for helping to purify the bloodstream, improve the energy capacity and impart a feeling of youthfulness.

**Licorice.** The long horizontal roots (rhizomes) are sweet and moist. The root enzymes have long been used to help improve the health of the bloodstream. The naturally sweet mucilaginous ingredient is a prime source of the emollient enzymes that help ease dry coughing and soothe bronchial disorders.

**Lobelia.** The entire herb plant may be used to help heal external wounds, sprains, and swellings. Enzymes help heal cuts and wounds when the lobelia herb is used as a natural poultice.

**Mullein.** Enzymes in the leaves and flowers have a pain-relieving benefit. It is said that these enzymes help nullify harmful bacterial infections and help promote youthful healing.

**Papaya.** The fruit, leaves and the juice are all prime sources of enzymes that reportedly help correct dyspepsia, poor digestion, and gastric catarrh. These enzymes also help soothe liver and stomach unrest.

**Parsley.** The entire plant is a source of healing enzymes. They have carminative (gas-dissolving) properties and also help soothe inflammations of the internal organs. The ripe seeds have often been used in pharmaceuticals to heal inflammation.

**Peppermint.** The leaves and flowering tops are rich in enzymes that help loosen accumulated mucus from the bronchial areas. They also help dissolve gas and act as a soothing digestive tonic.

**Psyllium.** Enzymes in the seeds have an emollient benefit in helping to remove accumulations from the bronchial-stomach-intestinal systems. Psyllium seeds often form the base for laxatives.

**Rauwolfia.** The root is a treasure of those enzymes which reportedly help cleanse the blood system and restore natural tranquility to the nervous system. Clinically, rauwolfia is used to help lower blood pressure and ease problems of hypertension. Ancients used the natural rauwolfia to promote a soothing relaxation and ease nervous disorders and headache.

**Rosemary.** The entire herb contains enzymes which use body nutrients to act as a natural headache tonic and boost sluggish circulation.

**Safflower.** The cold-pressed oil as well as the yellow and red portions of the flower are known for helping to lower cholesterol, to guard against arteriosclerosis and heart disease. Enzymes are abundant in the cold-pressed safflower seed oil.

**Saffron.** The dark orange to bright red blossom contains enzymes which are said to have carminative-diaphoretic (gas-dissolving and water-expelling) benefits that help cleanse the liver and bloodstream. Saffron enzymes have gentle digestive activators to alert a sluggish stomach.

**Sage.** The leaves have enzymes that are therapeutic to the kidneys, liver, gall bladder. Sage enzymes help cleanse these organs and promote a self-cleansing astringent action. Ancients used sage because its enzymes could heal scratches and wounds. Enzymes in sage leaves help wash out toxic wastes from the mucous membranes of the bronchial and alimentary tracts to help correct gastric problems.

**Sarsaparilla.** Enzymes in the roots help boost the health of the blood and skin. This herb is related to the ginseng and is reportedly beneficial in restoring a feeling of youthful virility.

**Sassafras.** The leaves and root bark contain enzymes that have gentle diuretic actions to promote the washing of the kidneys. The enzymes are said to help boost the health of the bloodstream. It is said that sassafras enzymes also help cleanse obstructions from the mucous linings and thereby promote better aeration of the body.

**Senna.** The pods are prime sources of enzymes that create a natural laxative action. The benefit here is that senna pods do not contain the resin that might cause stomach churning. Enzymes in senna pods act specifically on the lower bowel. This concentrated action makes this herb an all-natural laxative and bowel regulator.

**Spearmint.** The overground portion contains a mint-tasting enzyme that is helpful by dissolving gas bubbles, easing problems of catarrh and also easing spasms. Herbalists suggested spearmint for healing the bladder and kidneys and helping to bring about a natural diuretic action.

**Thyme.** The entire herb is a thriving source of enzymes that exert a therapeutic benefit to the bronchial structure. It is a mild stimulant and also a gas-dissolving herb. Ancients would burn thyme and inhale its fragrance as a means of self-cleansing.

**Vanilla.** The fully matured whole fruit of this herb is a reported top-notch source of enzymes that are quieting to the stomach and promote better water passage. It is believed that vanilla enzymes have a rejuvenating effect upon the brain, stimulate sluggish motor nerves and boost better metabolism. A bloodstream that carries oxygen better further helps improve emotional health.

**Watercress.** The entire herb is a succulent source of enzymes that speedily help cleanse the kidneys and liver. The enzymes use Vitamin C to help rebuild body cells and tissues, and keep the small blood vessels in a good condition. Enzymes then use other vitamins and minerals for enriching the bloodstream. Ancients prescribed watercress to correct "pale skin" or anemia. It was considered an herbal "blood tonic."

**Wintergreen.** The entire herb is a good source of those enzymes which help the body cleanse itself and also help promote better circulation. Wintergreen enzymes reportedly help correct problems of arthritis distress.

**Witch Hazel.** The leaves, twigs and barks of the stems have enzymes that offer an astringent-antiseptic quality to wounds and lacerations. Applied externally, witch hazel reportedly acts as a natural healing poultice for sore muscles, scratches, painful bruises and swellings. Enzymes in the herbal plant act as a natural sedative for such skin irritations as stings and burns. The leaves, when applied to the closed eyes, send soothing enzymes that reportedly ease tiredness and redness and help refresh the eyes.

## *Herb Enzymes—the Key to Youthful Health*

Herbal enzymes are Nature-created and are regarded the key to youthful health. The ancients worshipped herbs and used them as miracle healers. At the beginning of man's evolution, he had no books, college education, or scientific training. But he was given the gift of Nature. He became a student of Nature and observed how animals and birds selected plants for food and for healing. Slowly ancient man learned there were plants that nourished his system as well as plants that healed his system. Out of this beginning, the modern chemotherapeutic science of herbal enzymes developed. Today we know that enzymes are living substances that govern every body function. In herbal plants, we have a natural treasure of these life-giving enzymes. The ancients knew this wisdom and made use of it. Modern man is slowly rediscovering this wisdom and is turning to it as the hope for youthful health—at all ages.

*Highlights:*

1. In primitive days, witch doctors created miracles of rejuvenation with the use of herbs in secret rituals.

2. Herbal rejuvenation has evolved from superstition to modern scientific application. Modern medicines often contain herbal enzymes as ingredients.

3. Buy, use and store herbs with care to preserve their enzymes and healing properties. Use them *raw* for maximum enzymatic healing benefits.

4. A variety of different herbs are available to promote healing of almost every physical and mental disorder. Living enzymes abound in raw herb plants and are the keys to youthful health.

# 10

## *How the "Mono-Food Enzyme Program" Helps Rejuvenate Your Bloodstream for Enriched Youthful Health*

Digestive enzymes become rejuvenated under the influence of a traditional method of food combining known as the "Mono-Food Enzyme Program." It calls for eating only one kind of food at a time, or a combination of two such kinds of food at a time. Since most foods are divided into two categories–acid and alkaline–the enzymes needed to metabolize them may become diluted if there is a clash of improper combining. The bloodstream is in need of a steady supply of nutrients sent via an enzyme-healthy digestive system. The "Mono-Food Enzyme Program" reportedly creates a soothing digestive environment that is able to nourish the bloodstream and keep it sparkling clean and enriched. Since a healthy bloodstream is the key to a feeling of rejuvenation, it is important to consider the values of the "Mono-Food Enzyme Program."

### *Five Blood-Rejuvenating Benefits of the "Mono-Food Enzyme Program"*

Practical food combining creates better enzymatic harmony to offer these basic five benefits:

1. *Anabolism.* Enzymes work full force and without dilution in taking nutrients out of the foods and using them for the building and rebuilding of the red and white blood cells of the circulatory system.

2. *Assimilation.* Full-powered enzymes are able to transform digested nutrients into a vital part of the solids and fluids that enter into the bloodstream and other body systems.

3. *Catabolism.* Enzymes bring about the series of changes by which living matter or protoplasm breaks down into less complex and more stable substances within a blood cell or organism.

4. *Digestion.* Enzymes convert food into raw materials that are then used to nourish the entire body and to enrich and become assimilated by the bloodstream.

5. *Metabolism.* Enzymes work more youthfully under proper food combining to create food changes in the form of converting into components those essential substances needed for blood replenishment. Enzyme-favored nutrients can be broken down into molecules that can be utilized by the bloodstream for better health.

### Food Combining Creates Better Harmony

When ingested foods are combined properly, they create better enzyme harmony. The problem of fermentation caused by improper combinations may cause improper assimilation with the end result of a sluggish bloodstream. It is a simple and healthful practice to plan proper food combining for a few days a week to help give your enzyme system the ability to replenish itself and reward you with better health and a youthful bloodstream. The natural law states that what you eat is important, *but what you digest and assimilate adds to your health, strength and youthfulness.* Food combining helps invigorate the specific action of the digestive enzymes, which promote more careful timing of their secretions. Food combining focuses action of specific enzymes upon specific types of ingested foods.

*Special Benefit:* In observing the action of different enzymes, we realize that there is a special key to each lock: the ferment is the lock and its substrate the key. If the key does not fit exactly to the lock, a poor reaction occurs. An admixture of different types of fats, proteins and starches in the same meal may create an enzyme clash or diluting of the enzymes. The end result is poor assimilation of nutrients into the malnourished bloodstream. A more harmonious food combination is favorable to the enzymes so that they can work without interference of opposing forces in the digestive system.

The "Mono-Food Enzyme Program" enables digestive juices to work full force in their blood enriching tasks. *The effectiveness of the enzyme corresponds with the requirements of the food to be acted upon.* A simple blending of harmonious foods will help boost better enzyme vigor and youthful replenishment.

## *Ten Blood-Building "Mono-Food" Enzyme Programs*

1. *Eat Acids and Starches at Separate Meals.* Do not eat acid fruits with starch foods at the same meal. The ptyalin enzyme works more efficiently in an alkaline medium and is weakened or destroyed by acid at the same time. Fruit acids inhibit alkaline carbohydrate digestion and may cause fermentation. Nutrients are weakened and the bloodstream is denied the full vigor of the end products of foods. Such fruit acids as citric, malic and oxalic are antagonistic to the alkaline digestion of starches in the mouth and stomach. If food is acid at the start, there is weak hydrolysis (enzymatic digestion of starch) brought about by the saliva. This may lead to gaseous distress. A simple plan is to eat fruits about half an hour before starches so that enzymes can work full force to extract essential nutrients for body and blood rebuilding.

2. *Eat Proteins and Starches at Separate Meals.* Do not eat such protein foods as meat, eggs, and cheese at the same meal with starches such as potatoes, cakes, and bread. Protein enzymes act in opposition to starch enzymes. A mixture in the system may cause poor digestion. Undigested starch in large amounts in the stomach will absorb pepsin enzymes and thus inhibit the acid from entering into combination with the proteins to increase the free hydrochloric acid. The liver needs amino acids out of which plasma proteins (albumin, gamma globulin) enrich the bloodstream and build a natural immunity to infectious diseases. If protein digestion is inhibited because of the contrasting weakening starch diges-

tion at the same time, there is a reduced amount of amino acids made available to the bloodstream. The plasma proteins are weak and the bloodstream renders the person more susceptible to illness. Protein enzymes need an acid reaction; if there are too many starch enzymes creating an alkaline reaction, there is poor digestion. If meat and potatoes are eaten together, there is an outpouring of protein-digesting hydrochloric acid but this may weaken the outpouring of alkaline starch-digesting enzymes. Digestive disorder and "impoverished" blood may result. A simple adjustment of eating proteins and starches separately may be the key to blood enrichment through more potent enzymes.

3. *Eat One Protein Food at a Meal.* Two proteins of different characters and different compositions call for different types of digestive enzymes. These different enzymes vary in strength and character as they pour into the stomach at the same time. The valuable amino acids needed for body and blood replenishment may be diluted or weakened by improper enzyme clash. Milk protein requires a mild enzyme. Meat protein requires a strong enzyme. Egg protein requires a different timing in secretion from either milk or meat. Eaten together, enzymes struggle against one another since each one varies in power. This creates an indigestible food mixture. Poor assimilation often leads to "tired" or "pale" blood and constant susceptibility to colds and allergic infections. It is healthful to schedule meals so that you eat one protein food at a time for better enzyme harmony.

4. *Eat Proteins and Fats at Separate Meals.* Do not use meat, eggs, cheese, or nuts with cream, butter, or oil at the same meal. Fat depresses the action of the gastric enzymes and inhibits the pouring out of the proper gastic enzymes for meats, eggs, cheeses, nuts or other protein. The presence of ingested fat in the digestive system slows up the digestive enzymes. Fatty acids found in fat foods reduce the activity of the gastric enzymes, reduce the vigor of the enzymes that do get through, and lessen the amount of pepsin and hydrochloric acid in the stomach. *Suggestion:* Raw green vegetables are prime sources of healthful enzymes that supplement the body's own supply. If you must eat proteins and fats at the

same meal, begin with a large raw vegetable salad. The benefit here is that vegetable enzymes help counteract the inhibiting effect of fat so that protein can be metabolized. A proper combination will enable enzymes to take out Vitamins A, D, and E from fats to rebuild the blood cells and be sent via the circulation to organ depots as storage sites.

5. *Eat Proteins and Acids at Separate Meals.* Do not use meat, cheese, eggs, nuts with citrus fruits at the same meal. Acid fruits require a smaller amount of enzymatic flow than do protein foods. Fruit enzymes may inhibit protein enzymes. Furthermore, fruit enzymes are decidedly acid. If eaten at the same time with meats, it calls for an excess amount of acids in the system and this may be discomforting. The bloodstream needs a proper acid-alkaline balance and an outpouring of excessive acid may destroy the delicate red and white blood cells faster than they can be replenished. This may cause the person to become more susceptible to infectious illnesses. A simple program is to first eat the protein food and wait an hour or more before partaking of acid foods. It helps maintain better digestive balance and a more healthful bloodstream.

6. *Eat Sugars and Proteins at Separate Meals.* All sugars have an inhibiting reaction upon the secretion of digestive enzymes and upon the motility of the stomach itself. Sugars undergo no digestion in the mouth and stomach. They are digested in the intestine. When proteins are eaten at the same meal with sugars, the foods are delayed in the stomach for a prolonged period of time. While awaiting the completion of protein or starch digestion, they undergo fermentation. This sends excessive waste substances throughout the bloodstream. The accumulated sediment hampers a free-flowing circulatory system and anemia-like conditions may occur. A simple program is to eat natural sugars such as honey, sweet fruits, maple syrup, or molasses about an hour before a protein food. It creates better digestive harmony and reduces the possibility of fermentation in the bloodstream.

7. *Eat Sugars and Starches at Separate Meal.* Jellies, jam, sugar, honey, syrups, or molasses together with bread,

macaroni, spaghetti, pastries, potatoes at the same meal may lead to enzymatic fermentation. Starch digestion usually begins in the mouth and continues in the stomach. But sugars do not undergo any digestion in either the mouth or stomach but in the small intestine. When sugars and starches are eaten at the same time, they are delayed in the digestive system as they await metabolism of other foods. They tend to ferment very quickly under conditions of warmth and moisture in the stomach. This creates acid fermentation which is unhealthy for the bloodstream which is sent an outpouring of this corrosive liquid. Sour stomach is the first reaction. Anemia-like symptoms may later occur. A simple program is to eat natural sugars at a separate time from eating wholesome and natural starch foods. It helps promote better assimilation of nutrients for a healthy bloodstream.

8. *Eat One Concentrated Starch at a Meal.* It is healthful to limit the eating of starches. Overeating may lead to decomposition that upsets the delicate acid-alkaline balance of the body and the bloodstream which draws substances for nourishment from the digestive system. Starch-splitting enzymes vary: a potato reacts more favorably if there is no interference from macaroni enzymes. Furthermore, taking two or more starches at a meal may lead to overeating of this substance and cause an overloading of the bloodstream with the end products of fermentation. If two or more such starches are eaten at the same time, one or the other will be selected by enzymes for assimilation. The other starch must wait in the stomach where it undergoes fermentation with the creation of corrosive waste products in the system. The bloodstream struggles to wash out this sediment and health may be reduced because of this upheaval. It is simple and healthful to eat *natural* starch foods such as potatoes, whole grain bread, grains, or rice at a meal by themselves. Just *one* starch food helps send enzymes into the system to utilize the energy-building components that help boost the circulatory processes and provide a youthful bloodstream that flows on the spark of assimilated enzyme-treated starch foods.

9. *Eat Melons Alone.* Succulent fresh melons such as the watermelon, honeydew, casaba melon, cantaloupe, or musk-

melon undergo speedy decomposition in the stomach. When eaten alone, enzymes are able to devote full attention to them, and help metabolize their minerals into substances used by the bloodstream. But when eaten with other foods, enzymes are diverted and weakened. This swift decomposition means that fermented residues may end up floating in the bloodstream. Intestinal enzymes digest melons. But if melons are taken with other foods, they must wait in the stomach where they decompose, giving rise to gas and stomach discomforts. People who say that melons "disagree" with digestion would do well to eat them alone and without interference of other foods. It is a healthful way to promote better enzymatic assimilation of blood-building nutrients.

10. *Take Milk Alone.* Because of its protein and fat (cream) content, milk combines poorly with many foods. The first thing that occurs when milk enters the stomach is that it coagulates (forms curds). These curds tend to form around the particles of other food in the stomach, thus insulating them against enzyme juices. This inhibits their digestion for a long time until after the milk curd is digested. Milk does not digest in the stomach but in the duodenum (first part of the intestine). If milk curds gather in the stomach, other food enzymes are "sealed" in and cannot perform their digestive action. This slows up the entire metabolic process. Circulatory systems become sluggish and the bloodstream is denied its valuable supply of metabolized nutrients. Because milk is a gastric insulator, it should be taken alone and without the accompaniment of other foods. This helps send better nutrients to the bloodstream for overall better health.

## *Natural Foods Are More Easily Digested*

What Nature has combined, Nature can digest. Many foods are natural combinations of substances that have been especially created for natural digestion. What man has combined, Nature finds indigestible. *Nature never created a sandwich!* When you eat a Nature-created food that has a starch-protein combination, the digestive system can adjust its enzyme flow both as to strength and timing, to the digestive requirements of the food. But when two foods are eaten with different or *opposing* enzyme needs, the precise adaptation of enzymes to requirements becomes diluted. Satisfactory assimilation is slowed and the nutrients needed by the body and the bloodstream as carriers of health-building substances become unavailable. General health is subjected to a decline. A simple adjustment to healthful combining of natural foods will do much to create a healthful enzyme environment for overall health.

### *How a Simple Mono-Food Plan Enriched Marc's Bloodstream*

Marc O'H. ate wholesome foods including liver, eggs, fresh fruits and raw vegetables. But he had an iron deficiency. Marc was weak and easily tired. He had a sickish skin pallor. As an outdoor worker, he was frequently short of breath on the slightest exertion such as coming down from a work ladder. When he developed recurring headaches and dizzy spells, Marc O'H. realized that his health was declining. He was told that he had a form of iron deficiency anemia. He could not believe this because he always ate a balanced diet and emphasized iron-rich foods. But when he told about his exact menu, he explained that he liked to have a lot of fats with his meats. This was the problem. The fat-protein combination inhibited a free enzymatic flow that delayed or reduced the supply of iron to the bloodstream.

**Simple Adjustment Creates Youthful Bloodstream.** Marc was told not to eat any fats with his protein. This meant he was to avoid the use of butter, oil, fat, cream and milk when eating meats. This adjustment enabled his gastric glands to pour out a sufficient amount of enzymes that were *not* inhibited or diminished by the presence of fat. These enzymes could then metabolize the iron in meat for blood regeneration.

Originally, when Marc ate meat with a lot of fat, the iron formed complexes with other food substances. The iron became "bound" and could not be fully metabolized by enzymes for assimilation into the bloodstream. When iron and other nutrients gathered in the gastrointestinal tract where they were supposed to be metabolized, the presence of excess fat delayed this process.

Marc adjusted to eating a protein meat meal *without* added fat. Now the digestive enzymes could metabolize iron freely. They helped send this blood-building mineral through the mucosal cells of the intestine for absorption and transportation into the circulatory system. Enzymes further boosted Marc's basic health by stepping up red cell production and stabilizing the levels of plasma and body iron. It was this healthful balance that helped give Marc O'H. the enriched bloodstream he needed. Now he felt youthfully energetic. His skin was bright and colorful. He could climb up and down his work ladder with youthful agility and with a clear head and mind. Thanks to a simple Mono-Food Plan adjustment that enabled enzymes to feed iron to his bloodstream, life became well worth living.

## *How the Mono-Food Plan Helps Ease Acidosis*

Vivian A. was under constant pressure. She had to manage an active household. She was always on call from local social and school organizations. She had responsibilities as a part-time caterer. More and more obligations made her nervous and tense. The emotional reaction sent an excess of acid into her system. She was plagued by the burning sensation of gas bubbles trapped in her digestive system. Food became a lifeless lump. She was often wrenched over from the burning reaction.

**A Mono-Grain Plan Offers a Natural Alkaline Environment.** A nutrition-conscious naturalist friend suggested that Vivian A. eat grains *alone* as a means of creating a healthful alkaline environment. The plan was simple. *Four grains with a natural alkaline reaction are:* millet, rice, corn, and wild rice. These four alkaline-forming grains could be eaten during breakfast and lunch in as natural a form as possible. So that her digestive enzymes could extract electrolytes (elements) to form an alkaline environment, Vivian A. would not eat any opposing foods.

For mid-afternoon and dinner, Vivian A. would eat *four grains with a natural acid reaction:* oats, barley, rye, wheat. No other foods were to be taken, lest the enzymes become diluted or otherwise become diverted in their activities to help restore a healthful acid-alkaline balance.

*Results in Two Days:* Just two days of this Mono-Food Plan, using alkaline grains in the morning and acid grains in the evening, enabled her enzymes to send forth soothing substances into the bloodstream which then "washed" her digestive system. Soon a healthful acid-alkaline environment was achieved.

**Follows Mono-Diet Program.** This natural way to heal acidosis won over Vivian A. so that she tries to follow the Mono-Diet Program as much as possible in order to have a healthy digestive enzyme balance. She is free from acidosis and enjoys good food to help build good health.

### *Enzymes Create Blood-Building Substances*

When digestive enzymes are able to work without interference of conflicting foods, they help to create blood-building substances that enter into youthful regeneration.

Enzymes reduce food iron to a ferrous state. Enzymes now enter into transferring a portion of the iron to the plasma, while they store another portion in the mucosal cell to be released slowly into the plasma. Enzymes help the stored portion of iron become united with a special receptor protein complex known as *ferritin.* Enzymes now use this enzyme-created ferritin to facilitate absorption of iron from the intestines. Without enzymes, this portion of iron might be lost. Enzymes "know" that iron may be retained in the mucosal cells to the end of this life cycle (from seven to ten days) and then be cast off into the intestinal tract. Enzymes enter into this delicate precision mechanism to help enrich the bloodstream and keep it feeling youthful.

Enzymes also propel a considerable portion of absorbed iron into the plasma either by passing directly through the mucosal cells or after a temporary storage in ferritin bodies. Enzymes are found in plasma too. These plasma enzymes transport iron throughout the bloodstream.

Enzymes are able to quickly pick up and discharge iron, according to the need of the tissue it encounters. Enzymes send plasma iron to the bone marrow, the site of red cell manufacture. The bloodstream is nourished by the red cells to create a youthful circulation. Enzymes enter into almost all of these processes.

It is healthful to favor free-working enzyme activity by proper and practical food combining. This helps avoid fermentation and putrefaction of food and guards against gastrointestinal decomposition.

Enzymes need to work freely without the hindrance of unhealthy food combinations. Once the enzymes are nourished, they become "life forces" that are comparable to "batteries" that help charge up the reserve elements of the system to promote a youthful bloodstream for better health. The "Mono-Food Enzyme Program" helps nourish these "life forces" so that they can help regenerate and rejuvenate the rivers of life that flow within your body.

*In Review:*

1. The "Mono-Food Enzyme Program" is an all-natural key to bloodstream cleansing, replenishment, and rejuvenation.

2. The blood-building programs enable you to eat wholesomely with better digestive programs for better health.

3. Marc O'H. was able to enrich his bloodstream and enjoy more youthful vitality with a simple adjustment in his eating program. He did not combine fats with protein meals and gave his enzymes a renewed vigor of youthful regeneration.

4. Churning acidosis and trapped gas bubbles were problems that gave Vivian A. a burning stomach reaction. A Mono-Grain Plan gave her natural digestive enzymes that promoted a healing acid-alkaline balance and better health.

5. Practical food combining favors a free-working enzyme activity that promotes better iron absorption and a youthful bloodstream that promotes a feeling of vitality and energy.

# 11

## *Home-Grown Seed Sprouts: Secret Treasure of Youth-Building Enzymes*

The magic enzymes of youth are wrapped up in tiny growing seeds. These same miracles of self-rejuvenation are passed on to you when you use them right in your own home in the form of seed sprouts. You need not worry about the climate, problems of composting, poison sprays, bugs or weeds. You need no special equipment except an ordinary container and some water. There is no planting in soil, no plowing, no harvesting, no tedious gardening work, no long waiting time. You can grow seed sprouts in a glass jar, using nothing else except water and the seeds themselves. Seed sprouts will grow in the most primitive conditions, in any climate, can be planted almost any time of the year, require no soil, sunshine or waste in preparation. Furthermore, you can enjoy enzyme-rich seed sprouts within three to five days (sometimes sooner) after home growing in a glass jar—no long wait.

### *Seed Sprouts: Power-Packed Youth Enzymes in a Kernel*

Seed sprouts are treasures of youth enzymes in a kernel. It is believed that when seeds sprout, they multiply their natural nutrients until they become powerhouses of health- and youth-building substances. Let's see what seed sprouts contain to help you fortify your body with youth-building enzymes.

*Sprout Hormones:* Enzymes in seed sprouts create certain hormones called *auxins* which increase in size and number during the growing process. These sprout hormones are sparked by enzymes to act as organic catalysts to speed up the youth-building regenerative processes of billions of body cells and tissues. These enzymes then send the *auxins* into the glands out of which the youth-building hormones are then made.

*Sprout Proteins:* Enzymes in seed sprouts provide an excellent form of protein in its purest form. Cooking alters the complex structure of these molecules or destroys the supply of enzymes necessary for their utilization by the body, rendering them less valuable. Enzymes in seeds undergo an energizing process during sprouting and they take the protein from the plant and metabolize it within the kernel to a form of digestible amino acids that is considered to be far more valuable than animal protein.

*Sprout Vitamins:* Enzymes in seed sprouts take the raw materials from the plant and transform them into youth-building vitamins. These are pure, natural, organic vitamins that are untouched by chemicals, free from outside pollution. They are considered to be the purest vitamins available for your youth-building health. Some of these vitamins include:

Vitamin A–most whole grain cereal seeds have dormant amounts of Vitamin A that are increased many times during the enzyme process. This powerhouse vitamin is then sent by the enzymes into your body to promote better skin health and eyesight, and to assist in metabolism of various nutrients as well as essential fatty acids.

Vitamin B Complex–All members of the valuable B complex increase, sometimes as much as 1500 percent. Here are some examples. Whole grain cereal seeds contain enzymes that multiply the valuable Vitamin $B_1$ or *thiamine* content about 500 percent so they are then able to boost youthful energy in your system. In some seeds, Vitamin $B_2$ or *riboflavin* is increased close to 1350 percent. The enzymes then use this powerful vitamin to help your body metabolize other nutrients and also release energy to your billions of body cells. Oat seeds are especially powerful in this increase.

Whole wheat seeds contain *folic acid* which is multiplied five times by enzymes during the sprouting process. Mung beans (or chick pea seeds) contain *niacin* which is multiplied four times by enzymes during the sprouting process. Both of these B complex vitamins are needed to form body proteins, produce youthful energy, and make genetic materials for the cell nucleus, considered the "key" to youth. Enzymes multiply their value in a tiny all-natural seed kernel, uncontaminated by pollution.

Corn, barley and wheat seeds are prime sources of nearly all the B complex vitamins which are multiplied some 500 to 1000 percent during a few days of sprouting. Enzymes then use these super vitamins (including niacin, biotin, pantothenic acid, pyridoxine or Vitamin $B_6$, folic acid, inositol, thiamine and riboflavin for the building and rebuilding of the respiratory tract, in the formation of red blood cells, helping to rejuvenate the bloodstream and soothing the nervous system. All of these youth-building benefits are in easy-to-sprout home grown seeds.

Vitamin C–Enzymes multiply the Vitamin C content of seeds to make them a rich source of nourishment to help keep bone, teeth and blood vessels healthy. Enzyme-created sprout Vitamin C will also help form collagen, a protein that helps support body structures such as skin, bone and tendons.

One half cup of dry peas contains almost *no* Vitamin C, but after only four days of sprouting they contain 86 milligrams. One half cup of sprouted soybeans has as much Vitamin C as 12 cups of orange juice! Many vegetable seeds are vitamin-enriched by enzymes during the sprouting process, making them a natural and pure-as-Nature source of Vitamin C. Those who may dislike the tart taste of some citrus fruit juices will now discover a healthy and delicious way to get Vitamin C from crunchy, chewy seed sprouts.

Vitamin E–Enzymes use material within the kernel to create a powerful supply of Vitamin E, needed to prevent rancidity and oxygen depletion. Seed sprouts contain the natural forms of essential fatty acids that are nonfattening, unsaturated, nonhydrogenated, and cholesterol free! Seed sprout oils do not contribute to cholesterol buildup as do animal fats. Enzymes use the Vitamin E and the essential fatty acids in seeds to keep the skin and hair youthfully moist, to maintain water balance, and to help nourish the internal organs.

*Minerals:* Enzymes multiply the minerals within the seed so they can then help enrich your bloodstream, strengthen your bone structure, nourish your glands, and maintain a healthy water balance. Enzymes promote the increase of such minerals as calcium, phosphorus, iron, potassium, and magnesium.

Sprouted alfalfa contains nearly all the known vitamins and minerals as well as the fabulous chlorophyll. This is *a natural breath freshener* that is free from chemicals, synthetics, and additives. Alfalfa sprouts are Nature's own pure breath freshener. Just chew and eat them and feel the chlorophyll and the valuable minerals clear away the dry taste in your mouth and throat, leaving you as refreshed as a spring garden!

*Carbohydrates:* Enzymes perform a magic metabolism on the seed's supply of carbohydrates or starches. Enzymes change starch into easily digestible malt sugar. This makes the seed alive and bursting with as much as ten times the health and youth value it had when it was dormant. If you find ordinary starchy foods indigestible, you will enjoy the enzyme-predigested carbohydrates in seed sprouts. The benefit here is that seed sprouts are a rich source of enzyme-digested malt sugar. This substance then is propelled by enzymes into your bloodstream to create a "wake up and live" feeling that is brimming with youthful energy. It's the natural way to take your needed carbohydrates and enjoy vitality without starchy foods.

### *Seeds for Sprouting into the Elixir of Life*

Nearly any whole natural seed will sprout. Alfalfa (king of enzyme sprouts) usually is rated the best. Other good seeds are mung beans, soybeans, chick peas, garden peas, cress, black and red radish, chia, flax, broad beans, unhulled sesame seeds, safflower seeds, wheat seeds, rye, oats, corn, (almost all grains), fenugreek, buckwheat, lentils, and barley. Any organic and natural type of seed will give you an enzyme-enriched treasure of youth-building nutrients. You may use them singly or in combination for a smorgasbord of delightful good eating. The mixtures are endless. Seed sprouts are versatile as well as sources of youth-building enzyme nutrients.

*Where to Buy Seeds:* Any health food store or organic garden supply stores will carry them. Select seeds that are *not* chemically treated. Skip tomato seeds and potato sprouts as they are considered poisonous. All other organic and natural seeds are acceptable. Select any organic dried beans, peas, or grains for your sprouting.

*Where to Sprout Seeds:* Right in your own kitchen. No special place is needed for them. No need for sunshine either. Seed sprouting takes up as little space as any glass jar! A dark corner of a kitchen shelf is about all you need.

*What You Will Need:* In addition to organic seeds, all else you will need is a quart jar and some fresh water. You may want to use another method calling for a baking pan. Nothing else except, perhaps, a wire mesh and a dishcloth.

### *How to Sprout Seeds in a Glass Jar*

Here are the easy step-by-step instructions:

1. Carefully sort about half a cup of seeds. Discard any seeds that are broken or spoiled. Select only those that are clean, young and whole.

2. Wash the seeds thoroughly under free-flowing cold water.

3. Place the seeds in a large-mouth jar. The size should be one quart because the seeds will expand.

4. Fill with water and place in a cool dark place such as a corner of your kitchen shelf. Cover the top of the jar with a wire mesh or a layer of cheesecloth (or nylon netting). Fasten with two rubber bands. Let soak overnight or about 15 hours.

5. Next morning, drain water from the jar (through the wire mesh or cheesecloth) and rinse the seeds well. Now turn the jar upside down (no need to fill with water again!) and keep in a warm dark place.

6. Twice daily, rinse the sprouts by pouring lukewarm water into the jars, swirling to moisten the seeds and then pouring off the water.

7. The sprouts will measure from one-half to two inches in length within three days or sooner. When they reach this

length, you are ready to eat them and enjoy the "kitchen shelf-gardening" process that has enriched them with vitamins, minerals, proteins and energy-boosting enzymes.

### *The Glass Pan "No Work" Sprout Garden*

For a "no work" sprout garden to grow right in your kitchen, follow this easy step-by-step method:

1. Carefully sort about half a cup of seeds. As before, select only clean, young and whole organic seeds, discarding any that are broken or spoiled.
2. Wash the seeds thoroughly under free-flowing cold water.
3. Soak overnight in a large-mouth jar.
4. Next day, spread the seeds evenly and thinly on the bottom of a glass pan. Sprinkle generously with water two or three times a day. Cover the pan with a wet cheesecloth. Arrange the cheesecloth so that all sides hang over the edges of the pan.
5. Within four or five days, you will have a garden of sprouts in your glass pan. They are deliciously ready to eat.

### *How to Eat Your Sprouts*

Because enzymes have changed seed starch into more digestible natural sugar, sprouts can be eaten raw. They are speedily digested. The enzymes use the natural B complex vitamins to produce a form of energy that will make you feel alert and awake and youthful. This is why sprouts should be eaten raw. Since enzymes are destroyed by cooking, the secret of youthful vitality in sprouts is in their raw, unfired, untampered-with state of raw nature. For maximum benefits, eat sprouts raw. Some suggestions include:

*Enzyme Sprout Tonic:* Each morning, Gwen T. prepares a special Enzyme Sprout Tonic that she shares with her "too busy" salesman husband. She knows he scoffs at "bunny food" (vegetables) but she wants him to have energy and vigor in the morning. She herself needs youthful pep because she is a housewife and also part-time receptionist. So each morning, she puts half a cup of prepared sprouts into a blender. She adds half a cup of fresh fruit or vegetable juice. Then she whirs the blender for two minutes. Presto! She has a glass of all-liquid Enzyme Sprout Tonic.

Gwen T. drinks one glass for herself and gives another to her sleepy, grouchy husband.

**Enzymes Transform Husband into Cheery, Youthful Person.** Within a few moments, the enzymes from the sprouts send a powerhouse of dynamic nutrients shooting into her husband's bloodstream. He perks up. His eyes are alert. His face is clear. His brain is clear. He looks, acts, and feels youthful. The nutrients in the easy-to-make all-natural Enzyme Sprout Tonic have transformed him from a "morning grouch" into a "sunshine personality."

Gwen T. herself finds that this morning tonic gives her the youthful "go-go" power she needs in her double life as a housewife and office receptionist. It is Nature's miracle of youth-building enzymes in a simple kernel of seed.

*Soups:* Add a handful of sprouts to a bowl of any kind of soup just before serving. Do NOT cook the sprouts as this will destroy their valuable enzyme content. Just toss in the sprouts and enjoy with a healthy bowl of soup.

*Salads:* Add a handful of sprouts to any fruit or vegetable salads. Sprouts offer a crisp, crunchy joy to many fruit salads that might otherwise be ordinary. Sprouts add interesting texture to any salads. *Tip:* Add sprouts to jelled or molded fruit salads just before they start to solidify. You'll be adding a power-packed youth-building enzyme food that is every bit as delicious as nutritious.

*Main Dishes:* Add bean sprouts to stews, casseroles, and chow mein meals, and serve atop any cooked egg, meat or fish dish. The addition of bean sprouts adds living nourishment to cooked foods. This helps give you living vitality when eating cooked foods devoid of enzymes and other nutrients.

*Salad Dressing:* Blend tender sprouts with just a little fresh vegetable juice. When liquefied, pour over your salad. It gives you a young enzyme salad dressing that is brimming with nourishment as well as down-to-earth good taste.

*Breads:* Add ground wheat sprouts to breads and rolls. While the heating will deplete some of the enzymes, you will have a treasure of power-packed vitamins and minerals that help improve your youthful vitality. Another benefit is that these nutrients are then used by enzymes already in your system for regenerating your cells and tissues.

*Desserts:* Use any seed or bean sprouts on fruit desserts, or over any natural healthy dessert. This adds digestive-boosting enzymes at the end of a meal so they can now proceed to catalyze nutrients in your body to promote youthful rejuvenation.

**How to Store Sprouts.** While it is advisable to eat sprouts as soon as possible for their maximum enzyme "punch," it is permissible to store them in a closed jar in your refrigerator. They will keep a few days just like any other fresh vegetable, provided they are covered.

### *How the Swiss Use Sprouted Cereal Grains for Body Replenishment*

The secrets of Swiss healing are legendary. Many of their modern health spas rely upon the use of sprouted grains (also known as germinated grains) for boosting the enzyme and vitamin-mineral-protein content to provide a "dynamite" replenishment effect upon ailing or aging bodies.

At the famed Bircher-Benner health spa, a special "sprout meal" is given to tired and ailing persons. It is reported that this one meal, together with adequate rest, healthful foods and clean living, can remake and replenish "aging" bodies and give them a feeling of restoration of youth.

Here is the easy way that the Swiss use germinated cereal grains as an all-natural "youth food" that is brimming with enzymes for health restoration:

### *The Magic Swiss Youth Food*

*Grains:* Use rye, oats, barley or wheat grains. They should be whole grains, available at organic health food stores or from organic seed companies.

*First Day.* Evening: Wash the grains in a sieve under running water and put in a bowl. Cover with water. Keep at room temperature.

*Second Day.* Morning: Rinse and spread out to dry on a flat plate. Keep at room temperature.

*Third Day.* Morning: Rinse and spread out to dry on a flat plate. Evening: Place in a bowl and cover with water. Keep at room temperature.

By now, the grains should have developed up to one-inch-long sprouts. They are ready to eat.

**Magic Enzyme Meal.** Mix the sprouted grains with bananas, honey and some fruit juice. Eat as a meal.

**How the Magic Enzyme Meal Restored Erwin's Digestive System.** As a production manager, Erwin V. was always under tension. His digestive system was always churning. Everything he ate felt "tied up in knots." He would experience stabbing pains in his stomach after many a heavy meal. He wanted to go to a health spa to recuperate but his finances were limited. He met several other folks with the same stomach churning distress who were reportedly healed at a Swiss health spa. They would eat the simple Magic Enzyme Meal at least once a day. Erwin V. decided to try it. He made the sprouts in his kitchen, using a mixture of organic grains. Then he mixed the sprouted grains with bananas, honey and some fruit juice. He ate this as a meal.

*Benefits:* The tremendous store of nutrients as well as the dynamic enzymes immediately boosted the effectiveness of his sluggish digestive system. Furthermore, the enzymes took the fructose from the honey and the natural carbohydrates from the banana and transformed them into usable energy-producing substances. These were able to "energize" his digestive-intestinal organs so that Erwin V. could digest and assimilate what he ate. Heretofore, he complained that it was a "lead weight" in his stomach. Thanks to the enzymes in the Magic Enzyme Meal from Switzerland, Erwin V. was able to metabolize his foods. Soon his digestive system felt clear, youthful and sparkling clean. He boasted that he had a "young" stomach—thanks to the miracle enzyme power of sprouted grains and to Swiss secrets of healing.

**Revives, Rejuvenates, Regenerates.** Germinating the cereal grains the Swiss way is an Alpine secret of promoting the youth-building powers of the plant. Enzymes are increased a thousandfold. Vitamins and minerals are boosted some five hundred percent. *Most important* is that sprouting enhances the unsaturated fatty acid content of the grains. Three essential fatty acids (unofficially known as Vitamin F), arachidonic, linoleic and linolenic acids multiply at a phenomenally speedy rate under sprouting. These three essential fatty acids (called essential because the body cannot make them and they must be obtained from foods) are needed for regenerating cells and tissues, and for acting as carriers of Vitamins A, D and E to the body's glands and organs. Enzymes propel these three essential fatty acids. Enzymes also deposit the vitamins in the lifestream to help revive, rejuvenate and regenerate the body from top to toe.

The enzymes in cereal grains are powerhouses of these nutrients. The cereal grains, when sprouted, contain more than just enzymes, vitamins, minerals and forms of proteins. They also contain the three essential fatty acids. All work in harmony, propelled by enzymes to promote this miracle rejuvenation effect.

Small wonder that this Swiss healing "secret" has worked wonders for so many. Today you can make the same cereal sprouts right in your own home. A little bit of effort for a feeling of lifetime youth!

### *How Orientals Use Bean Sprouts for Eternal Youth*

The Orientals are a hardy people. They must devise methods of survival and long life under difficult conditions. Much of the land in the Orient can be utilized solely for rice. Cattle are difficult to sustain upon rocky plains with sparse vegetation. Beans appear to grow everywhere. The wise Orientals have learned to multiply the food value of most beans by sprouting them. Even soybeans, which contain almost all the known nutrients needed to sustain life, have been improved by the Orientals to increase their youth-building properties many times over.

Bean sprouts, first originated in the Orient, are regarded as the "fruit of eternal youth." For many Orientals, the bean sprouts have done more than serve as nourishment in times of famine and deprivation. The bean sprouts have given them a treasure of enzymes and youth-building nutrients so that they could withstand plague, devastation, starvation and travail. The sprouts appear to have given them eternal youth. Bean sprouts may well hold the key to the mysterious secret of "eternal youth" among the Orientals.

### *The Eight-Step Oriental Way to Sprout Beans*

1. Select a stock of clean bright new organic beans. Yellow seeds have more transparent skins than the black beans, but both are used by the Orientals.

2. Select only those beans that are fresh, clean and whole. Discard all that are broken or spoiled.

3. Wash the beans under fresh cold water. Now place them in a one- or two-quart glass jar. (*Note:* Bean sprouts increase to more than five times their original size.)

4. Cover the seeds with four times their volume of lukewarm water. Let them stand overnight until they grow in volume.

5. Next morning, pour off the water. Wash and rinse the swollen beans.

6. Cover the top of the jar with a piece of cheesecloth or any thin cloth. Fasten securely.

7. Tilt the jar so that excess water can drain away. Place it in any dark place (such as a cupboard).

8. Every four hours, rinse the sprouts with cool water. Pour off the water. The benefit here is that you can wash away any decay that may have developed. After rinsing, return the jar to the dark corner in its slanting position.

**Ready Within Five Days.** Within five days, you should have fully grown bean sprouts. They should measure as much as two inches in length. The Orientals now wash them to free them from loose hulls or seed coats. Now you can use them as a vegetable salad or other good eating.

**Soybean Sprouts.** Orientals prize these above all other sprouts. They have more enzyme and nutrient power than any of the others. You may want to use these more frequently.

**Mung Bean Sprouts.** These are the tiny green beans out of which the bean sprouts used in chop suey dishes are prepared. They offer a powerhouse of dynamic youth-building enzymes and assorted nutrients.

**String Bean Sprouts.** They are believed to create a healthy alkaline reaction in the system. They are rich in potassium, phosphorus, calcium, magnesium and sulphur. They have good natural enzyme-digested carbohydrates as well as goodly amounts of proteins. The enzymes in string bean sprouts take all of these nutrients and use them for the building and rebuilding of your body.

The Orientals have developed "bean sprout culture" to a new peak. Today sprouts are recognized as possessing almost all of the elements needed to sustain life. The modern Oriental calls it "Magic Youth Food" and his record of longevity and good health despite generations of strife and starvation may well attest to his wisdom. Indeed, the Oriental bean sprouts are "magic youth foods" with their treasure of life-sustaining elements–in the absence of other foods! Oriental bean sprouts, such as you can make at home, may be the long-sought-after "perfect food." It is–to the Oriental.

### *How a Bus Driver Uses Sprouts as "Complete" Nourishment*

Michael B. has a difficult route as a bus driver. He also works erratic hours. This plays havoc with his mealtimes. Often he must take a sandwich and a cup of coffee with him that he eats and drinks either at a rest stop or while driving.

A problem is that this "dead" food deprived Michael B. of needed nourishment. Deficient in enzymes, his reflexes started to swerve. Soon he could not see clearly during night driving. He could not see traffic signals or signs. He had a few near accidents. He was a peril to himself and his passengers.

**Simple Food Change Boosts Youthful Reflexes.** An Oriental passenger friend once suggested that Michael B. take along a jar of sprouts. All Michael B. had to do was to prepare a good supply of home-grown sprouts in advance. He would store them (covered, in a glass jar) in the refrigerator, as with any vegetable. Then he would take a portion with him for a hearty meal in itself. Yes, a half quart of sprouts with a cup or two of herbal tea or coffee substitute could constitute a meal. Certainly it was more nourishing than his bleached, pasty bread sandwich and sugary coffee!

*Benefits:* The enzymes in the assorted sprouts took the essential fatty acids to promote a feeling of satiety to Michael's system. The enzymes then took the multiplied vitamins, minerals and proteins to help nourish his body and send streams of oxygen-carrying blood to his brain for quick alertness. His reflexes were youthful. Now he could drive long hours under difficult conditions, thanks to the Oriental method of using home-grown sprouts as a "perfect food" for youth and health.

**Dynamic Health in a Tiny Seed.** Wrapped within its protective covering, a tiny organic seed carries hope for dynamic health. Sprouting increases the vitamin, mineral, protein, enzyme and other nutrient value so many times that many regard sprouts as the key to eternal youth.

Sprouts help you maintain biological youth. The source of dynamic enzyme rejuvenation is that you eat the *entire* sprout. The best way to obtain enzyme rejuvenation is to consume a vegetable in its entirety–with its structural parts and leaves, right down to its base. A plant is a living substance. Life means youthful movement. Throughout the sprout's life there is a cyclic flow of energy that has a high vibration rate. If any segment of the plant is severed from its main body, there is a depletion of fluid and energy similar to a severed artery. Therefore when you eat the entire plant (such as a sprout), you are eating the whole plant or the full flow of cyclic energy. This gives you the youth-giving enzymes that nourish the plant and now serve to nourish you.

A growing plant (such as a sprout) is a living plant. Once a plant such as any fruit or vegetable has been cut from the soil, it loses its living juices. Not so with a growing, living sprout. Just eat the entire sprout and the effect is the same as eating a *living plant* that is a source of dynamic health and youth. No other plant–except the sprout–can give you that enzyme power. No other plant can be eaten alive, since being picked means it loses its life–except the sprout. It is Nature's own way of giving you the magic enzymes needed for a feeling of eternal youth and eternal health.

*Summary:*

1. Seeds are filled with enzymes and nutrients needed to sustain life and promote youthful vitality. Sprouting (germinating) right in your own kitchen will cause an "energy explosion" from within the seed to give it a treasure chest of youth-building nutrients.

2. Sprouting increases the sprout's supply of enzymes, hormones, proteins, vitamins, and minerals. It also predigests carbohydrates into "instant energy" malt sugar to give you a "wake up and live" feeling.

3. Sprout seeds right in your own home. Your only needed equipment includes organic seeds or beans, a glass jar, some water, and a mesh cloth covering. No soil, sunshine, plowing, harvesting, sprays, insecticides, fear of fallout, or climate problems. Grow enzyme-rich sprouts without any difficulties.

4. Gwen T. and her husband feel youthful vitality from a delicious easy-to-make Enzyme Sprout Tonic.

5. For "dynamite" replenishment, try the Swiss secret: a Magic Swiss Youth Food—a Magic Enzyme Meal. This helped Erwin V. rejuvenate his "knotted" stomach and sluggish digestive system.

6. Orientals consider bean sprouts (instructions given) as a "perfect food." Follow the easy eight-step Oriental way to sprout beans for better health. Oriental bean sprouts gave bus driver Michael B. the youthful reflexes he needed on his tiresome route. Sprouts are Nature's sources of dynamic health in a tiny seed: Nature's magic enzyme plant!

# 12

## *How to Use Enzymes to Revive Your Glands and Boost Your Supply of Youth-Creating Hormones*

Your glands are the gateways to youth and health.

To open these gateways, you need enzymes. These are the catalysts that bring nourishment to the gateways, feed your glands, and help alert them so that they are then able to send forth a stream of youth-creating hormones throughout your body.

Enzymes then enter into the composition of the hormones themselves. They boost their quantity, strengthen their flowing power, propel these living rivers throughout your body from head to toe. Without enzymes, ingested nutrients would remain (along with eaten food) as a solid mass. Your body functions would come to a total standstill. Your glands would remain inert. The gateways would be closed. Your hormones would become dead lakes. Your health would be in peril.

**Enzymes Control Health of Glands.** A youthfully healthy body requires youthfully healthy glands which is possible by an adequate and abundant supply of enzymes from natural raw foods. Enzymes act as spark plugs that trigger and stimulate your glands into secreting youth-building hormones to rejuvenate your feeling of well-being, both physically and mentally. Enzymes stimulate these gateways into sending forth hormones and other secretions that promote a healthy and youthful functioning of just about all of your body parts from top to bottom, from the inside to the outside. Your health depends upon an adequate supply of raw food enzymes. Your glands are waiting for these enzymes to give them the impetus and nourishment to send forth a never-ceasing supply of youth-creating hormones. Let's see how this works and how you can use simple, everyday foods and special tonics to nourish your gateways and reap the rewards of "hormonization" through natural enzymes.

**How Glands Hold Key to Youth.** A gland is an organ that forms and releases (secretes) substances that promote a regeneration benefit to a related body function. None of the body glands act independently of the others. There are feedback mechanisms between each of them. If there is a deficiency in a secretion of one gland, there may be a deficiency in the secretion of another gland. Or if there is an excess in the secretion of one gland, there may be an excess in the secretion of another gland. Often there is a seesaw imbalance so that if one gland secretes *too much,* another gland may conversely secrete *too little.* This may adversely affect your youthful health. So we see that a well-balanced glandular system does hold the key to youth.

**How Hormones Boost Youthful Vigor.** A hormone is a substance that is secreted by a gland. The name hormone is taken from the Greek *hormon,* meaning to "stir up" or to "excite." A hormone is propelled by enzymes into the bloodstream to regulate the functions of other hormones. Hormones circulate in almost all body fluids to promote youth-regenerating functions of nearly all the body processes and mechanisms. Enzymes take these hormones and use them to promote better digestion, to keep the heart pumping healthfully, to boost the flow of a youthful bloodstream. Enzymes use hormones for transporting more oxygen and nutrients to muscle cells to provide vigor and strength to meet the challenges and responsibilities of daily living. An enzyme deficiency correspondingly means a sluggish hormone flow and a subsequent letdown in physical energy and emotional health. So we can readily appreciate the value of enzyme-nourished hormones that flow through the open gateways and promote health from head to toe.

Now let's see how you can use natural raw food enzymes to help revive and boost the health of your glands and hormones.

### A Magic Enzyme Program for the Pituitary

*Gland Location:* Comprised of an anterior and posterior lobe, it hangs like a berry from the underside of the brain, occupying a small hollow in the floor of the skull, and is considered the "master gland" of the body.

*Hormone Function:* Helps maintain smooth muscles, regulates water balance, and helps control kidney function. Also stimulates the adrenal cortex to secrete ACTH, a hormone known for helping to heal problems of arthritis. It is an important growth hormone too, and helps keep the body in youthful vigor and strength.

*Magic Enzyme Program:* Essential fatty acids as well as Vitamin E are natural propellants that are used by enzymes to stimulate the pituitary to send forth its needed hormones. Enzymes use these substances to protect the pituitary hormones, to prevent their destruction by conserving oxygen, and to invigorate them to perform adequately. *Program:* Daily, take up to four tablespoons of cold-pressed wheat germ oil, safflower seed oil, or sunflower seed oil. This is your source of enzyme-packed Vitamin E and the essential fatty acids that are then used to nourish the pituitary and become part of the hormones it will then send forth. The cold-pressed, nonprocessed, unhydrogenated oils are prime sources of Vitamin E as well as the necessary essential fatty acids that are used as food by the gland. Your enzymes need these oils to nourish the gland. A simple yet reportedly effective program.

### A Magic Enzyme Program for the Adrenals

*Gland Location:* Consisting of two parts, the medulla (central portion) and the cortex (covering), this is a small set of glands located just in front of the kidneys.

*Hormone Function:* When the body is subjected to emotional and/or physical stress, the glands send forth two hormones, namely epinephrine (adrenaline) and norepinephrine. These hormones increase the heartbeat rate to meet the challenges. They increase the supply of blood sugar under stress conditions and boost the metabolic rate. The norepinephrine transmits messages through the nervous system in conjunction with the complex mechanism involved in stress situations created by pain, shock or such adverse emotions as anger or fear. The adrenal glands also need enzymes to manufacture *cortisol,* according to the direction of the pituitary gland, on the basis of the ACTH that the pituitary sends to the adrenals. Enzymes use cortisol as a hormone for metabolizing carbohydrates to produce the energy and protein which build and rebuild body cells, tissues and organs. Enzymes also stimulate the production of the aldosterone hormone which you need for the regulation of potassium and the retention of sodium by way of the kidneys.

*Magic Enzyme Program:* The B complex vitamins as well as Vitamin C are needed by enzymes to nourish the adrenal glands. These nutrients are stored abundantly in the adrenals, perhaps more than in any other body part. A deficiency of either of these nutrients may cause the adrenals to go awry, sending forth an increased supply of hormones; this may cause exhaustion of the gland's secretion and create a body shortage. So to enjoy good health, feed nutrients to your adrenals. *Program:* Mix two tablespoons of brewer's yeast powder (sold at many health food stores and herbal pharmacies) in a glass of fresh citrus fruit juice. Orange and grapefruit juices are highest in Vitamin C content. Stir vigorously. Drink this as a tonic in mid-afternoon. You will find that the high-powered B complex in the yeast will be used by enzymes to become mixed with the Vitamin C of the citrus juice. The enzymes will then take this treasure of nutrition to the adrenals and nourish them so they can send forth a healthy and steady supply of necessary hormones to maintain body health. Without these nutrients, your enzymes are helpless since they lack the raw materials out of which adrenal gland hormones are made.

### *A Magic Enzyme Program for the Thyroid*

*Gland Location:* A butterfly-shaped gland that rests against the front of the windpipe. It consists of two lobes, one at either side of the Adam's apple, joined at the lower ends by a bridge of glandular tissue across the upper part of the trachea (the windpipe, running from the voice box down to the bronchi of the lungs).

*Hormone Function:* Enzymes help the gland secrete thyroxin, a hormone that stimulates the activity or metabolism of body tissues and cells. Too much thyroxin (hyperthyroidism) speeds up metabolism. Too little thyroxin (hypothyroidism) slows it down. This causes irregular weight, erratic pulse, poor breathing, problems of goiter, bulging eyes, and nervous disorders. Thyroxin is stimulated by enzymes to promote the release of energy in the tissues from combustion of glucose. This helps give the body and the mind the working materials needed for physical and emotional youthfulness.

*Magic Enzyme Program:* A weak or lazy thyroid can turn a young person into an older one. A well-nourished thyroid that receives an abundance of enzyme-propelled nutrients will help keep this gland in youthful movement. The thyroid requires iodine and the B complex vitamins, especially Vitamin $B_1$. Your enzymes must have these nutrients or else your thyroid may become starved and "hysterical" and perform at a rapid or sluggish pace. A balance is needed. Your enzymes can create that youth-building balance with iodine and the B complex nutrients. *Program:* Kelp is a powdered sea salt sold at most health stores and herbal pharmacies. This is a powerhouse of minerals as well as iodine that are used by enzymes for nourishing the thyroid. The B complex vitamins are very abundant in natural brown rice (available at almost every food market–but be sure the label specifies it is *natural brown rice.* Avoid the bleached white rice which is chemically treated (chemicals are antagonistic and destructive to enzymes) and devoid of natural vitamins. Daily, sprinkle kelp powder (a half teaspoon gives your enzymes the needed iodine) over meals in place of salt. Sprinkle a bit of kelp over a bowl of steaming natural brown rice. This gives you a double-barrelled supply of both iodine and the needed B complex vitamins. Your enzymes will use these nutrients to nourish your thyroid so that it will dispel fatigue, ease lethargy, warm up cold fingers and toes, boost your healthy sexual interest and also help control weight. *A magic enzyme program that is found in two simple, everyday all-natural foods that cost pennies per serving–but offer a pot of gold of youthful health!*

### A Magic Enzyme Program for the Parathyroids

*Gland Location:* Four tiny glands, each about the size of a pea, are located at the four corners of the thyroid, in the neck.

*Hormone Function:* The gland produces the *parathormone* which helps regulate the body's use of calcium and phosphorus. Enzymes are needed to boost this hormone's function in regulating the mineral levels in the bloodstream. An enzyme deficiency may cause mineral imbalance, overexcitability of muscle and nerve, twitching, cramp and tingling sensations. If the gland does not have sufficient enzymes, it goes wild by withdrawing too much calcium from the bones which then become fragile. An excess of this hormone (caused by enzyme deficiency) may lead to calcium deposits in other than bone tissues, as in circus freaks said to be "turning to stone." More serious symptoms may be convulsions, cataracts, hearing problems, loss of appetite and loss of hair. Your parathyroids need an adequate amount of nutrient-carrying enzymes to help keep your body in youthful health.

*Magic Enzyme Program:* To help keep your parathyroids adequately nourished, they need minerals. But more important, they need raw food enzymes that can propel these minerals to maintain a healthy glandular balance and healthy hormones. A wholesome and tasty food that is sold almost everywhere for a tiny cost can give you this double-action enzyme-mineral hormone food. That simple food is yogurt. *Program:* A cup of plain yogurt is a treasure of living enzymes as well as much-needed calcium-phosphorus. The enzymes take these minerals and propel them to the parathyroids which use them for the manufacturing of a steady supply of youth-building parathormones. The enzymes are needed to metabolize the calcium and phosphorus so they can be used by the glands. That is why simple yogurt is so effective as a gland food. It is a prime source of both enzymes and the needed minerals to create this magic effect in rejuvenating the glands and subsequently rejuvenating the body. *Suggestion:* A cup of yogurt eaten at breakfast time is a good way to enzyme-energize your glands in the morning. Also, have yogurt during the day to perk up sluggish glands. It's the natural way to enzyme-enrich your parathyroids with needed minerals.

### *A Magic Enzyme Program for the Pancreas*

*Gland Location:* A long, slender organ that is located near the stomach. It is a dual-purpose organ. It is an offshoot of the intestine and is also a digestive gland. That is, the pancreas is part of the digestive system and secretes pancreatic juice, which enzymes release into the intestines. Enzymes now use pancreatic juices to help break down food for assimilation. Without enzymes, food would remain a solid mass in the stomach.

*Hormone Function:* The pancreas uses enzymes to secrete hormones (insulin and glycogen) which metabolize carbohydrates and sugar.

*Magic Enzyme Program:* The pancreas must have enzymes to issue insulin to promote the uptake of sugar from the blood by the tissues for storage and combustion. The enzymes use glycogen to promote the release of sugar from the tissues (mainly the liver) into the blood. An enzyme deficiency may cause a sluggish pancreas, an inadequate amount of insulin, an oversupply of undigested carbohydrates and a predisposition toward diabetes. Enzymes are secreted by a healthy pancreas to send forth insulin, the spark that ignites the fire for burning sugars and starches and transforming them into youthful energy.

*Program:* Curtail intake of chemically treated and devitalized sugar and starch foods. Switch to doctor-approved whole grain breads and cereals. In particular, you need to send protein-carrying enzymes to your pancreas. This gland needs the amino acids from metabolized protein for creating of healthy insulin and other sugar-burning hormones. Enzymes bring about this internal combustion process. A good source of top-notch protein is in a product available just about everywhere. It is unflavored gelatin powder!

It is a sugar-free food, very low in calories and carbohydrates, and also a prime source of the needed amino acids. Enzymes will take up these gelatin powder amino acids and use them in building and maintaining the health of the pancreas. You can obtain unflavored gelatin at almost any food market and health store.

*Suggestions:* Here's how to give your pancreas a powerhouse of enzyme amino acids. (1) Mix one envelope or one teaspoon of unflavored gelatin into fruit juice, vegetable juice or skim milk. Stir briskly and drink immediately. (2) Mix one envelope of unflavored gelatin in half a cup of cold water to soften. Add half a cup tomato juice. Stir and drink promptly. (3) Sprinkle one envelope of unflavored gelatin on 1/4 cup cold water. Add 3/4 cup hot broth or a bouillon cube. Stir until gelatin and bouillon are dissolved. Makes a good warm drink. *Benefits:* Enzymes in the juices combine with enzymes in the gelatin and take up the amino acid content. Within moments after drinking *any* of these tonics, the enzymes are shooting forth amino acids to the pancreas for healthy hormones and youthful health. Simple, inexpensive, but a gold mine of natural hormone health.

### *A Magic Enzyme Program for the Thymus*

*Gland Location:* Situated on the windpipe below the thyroid in the root of the neck.

*Hormone Function:* Stores forms of proteins that are taken by enzymes for use by the adrenals in times of stress or tension. The gland also contains many developing lymphocytes or white blood cells which are taken by the enzymes to be distributed throughout the bloodstream to help promote immunity and guard against infection.

*Magic Enzyme Program:* Fresh raw fruits and vegetables are the prime source of enzymes which are then able to nourish the body and be used by the thymus for the manufacturing of more white blood cells. *Program:* Daily, eat a large raw fruit salad as well as a large raw vegetable salad. You will also be benefiting your thymus gland with a shower of revitalizing enzymes by eating a complete raw vegetable meal as a main dish. The enzymes will then speedily revitalize the thymus to send forth hormones that will carry white blood cells to create a form of immunity to infectious microbes. Folks troubled with allergies and frequent colds would do well to eat heartily of raw fruits and vegetables. This will give your thymus the needed white blood cells that your body, in turn, uses to fight off infectious microbes and allergens. It's the natural way to help build immunity.

### *A Magic Enzyme Program for the Male Sex Glands*

*Gland Location:* The testes are the primary sex glands of the male, and are situated in the groin area.

*Hormone Function:* Enzymes help create testosterone, a male sex hormone that induces and maintains important sex characteristics. This hormone also promotes muscle resiliency, stimulates growth, strengthens porous bones and appears to influence sexual vigor. Enzymes need testosterone in order to help give the male a healthy libido.

*Magic Enzyme Program:* Enzymes require a number of vitamins and minerals to invigorate the testes to create various hormones. Enzymes trigger the entire glandular network for harmonious flow of such youth-virility hormones. Enzymes help promote androgens, the masculine sex hormones. The androgens are influential in building a healthy virile urge in the male. Enzymes induce hormones to regulate sexual metabolism; the processes in living cells by which new spermatic fluid is assimilated within the system to promote healthy libidinal functions.

Enzymes also use these hormones to help bring about a youthful growth of cells and connective fibers and build resistance to the degenerative cell processes that accompany problems of male weakness and premature aging. *Program:* One mineral needed by enzymes is zinc. It increases the activity of some 25 enzymes essential in hormonal metabolism. Zinc binds the enzyme to the substrate it acts upon. Zinc is found in natural brown rice, raw fruits and vegetables and nonprocessed pumpkin seeds, sunflower seeds and most nuts. Eat a handful of these seeds and nuts daily for boosting the zinc content. *Special Enzyme-Zinc Tonic:* In one glass of any fresh vegetable juice, mix two tablespoons of desiccated liver, two tablespoons of brewer's yeast and one teaspoon of wheat germ flakes. Stir vigorously. Drink one glass in the morning and a second glass in the early evening. You'll be sending a stream of enzymes and zinc together with vitamins and minerals throughout your system to energize your sex glands to help promote healthy vigor.

### A Magic Enzyme Program for the Female Sex Glands

*Gland Location:* The main female sex glands are the ovaries. These are two small almond-shaped glands set in the pelvic cavity (lower abdomen) just below the navel. One is about three inches to the right and the other the same distance to the left of the center of the body.

*Hormone Function:* Enzymes stimulate the ovaries to issue forth the elaborate female sex hormones that help keep a woman's skin soft and moist, give her a youthful libidinal urge, and govern the other glands. All work together to give her a feminine and pleasant appearance of body and mind. Enzymes act as the spark plugs to promote the distribution of these valuable female hormones.

*Magic Enzyme Program:* Enzymes in raw fruits, vegetables, seeds and nuts promote the flow of estrogens, the female hormones. These enzymes promote a healthful issuance of such hormones that regulate a woman's sexual-reproductive metabolism and help keep her young in mind and body. *Program:* Daily, eat fresh fruits and vegetables. Drink lots of fresh raw fruit and vegetable juices. Eat sunflower and sesame seeds and healthful nuts. They are prime sources of those enzymes needed by the glands to promote youth-building hormones.

### *How a Simple "Sex-Enzyme" Program Boosted Virility*

Andrew C. was in his middle forties, but his frustrated wife often said that he was in his eighties insofar as being a sexy young husband was concerned. Andrew C. had a high basal metabolism rate which reportedly depleted his nutrient supply and took its toll upon his enzymes. The adverse reaction was in his sluggish hormonal flow and reduced virility. He followed a very simple "Sex-Enzyme" Program with all-natural foods to help build and rebuild his glands so they could give him "forever young" hormones. Here is the program followed by Andrew C.:

1. *Vitamin A for Enzyme-Boosting Hormones.* This vitamin helps send enzymes to the glands so they can make sex hormones or steroids. Enzymes use Vitamin A for taking these hormones and sending them throughout the body to promote youthful health. A deficiency of Vitamin A reportedly may cause impotence. Isobel W. Jennings, M.D.,[1] says, "Advanced Vitamin A deficiency in the male rat is

---

1. Isobel W. Jennings, M.D., *Vitamins in Endocrine Metabolism,* Springfield, Illinois: Charles C. Thomas Publishing Co., 1970.

equivalent virtually to castration and is associated with degenerative changes in the germinal epithelium of the testes (tissues covering the testes within the scrotum) and the production of abnormal spermatozoa." The doctor notes that "Vitamin A deficiency interferes with the synthesis or release of androgens (male hormones)." *Program:* Andrew C. took two tablespoons of fish liver oil daily for a powerhouse of enzyme-needed Vitamin A. He also ate lots of green leafy vegetables and vegetable oils for more Vitamin A.

2. *Vitamin B Complex for Enzyme Vigor.* Andrew C. took Vitamin B complex in the form of whole grain unbleached breads and cereals. He ate lots of desiccated liver. He sprinkled brewer's yeast flakes over his whole grain cereals. This gave him the B complex vitamins, especially niacin and folic acid which transform substances into needed sex hormones. The *enzymes* in these whole grain foods promoted this transformation.

3. *Vitamin C for Hormone Production.* Daily, Andrew C. would drink fresh fruit juices. He would eat lots of fresh fruits and raw fresh vegetables. This gave him lots of powerful enzymes that took the same Vitamin C in the foods and used them to stimulate the tubules of his testes to produce needed sex hormones. A natural way to give his body the needed youth-building hormone and provide mutually satisfactory sex relations in his marriage.

4. *Raw Seeds and Nuts for Vitamin E.* Enzymes need Vitamin E to protect the sex hormones and to enter into the intricate functions involved with manufacturing the needed fluids. Andrew C. would eat lots of raw seeds and nuts for a good supply of these valuable enzymes together with Vitamin E. Enzymes also used the unsaturated fatty acids from the raw seeds and nuts to help lubricate the sex glands and to protect against premature sexual senility through hardening of the arteries. He also used cold-pressed unrefined oils such as corn, safflower and sunflower oils for more Vitamin E intake. His glands absorbed the Vitamin E and the enzymes metabolized them to give him the vigor that made him look, feel and (most important) act young again!

5. *Minerals and Proteins for Enzyme Balance.* To round out and balance his enzyme program, Andrew C. ate whole-

some and natural foods for minerals and proteins needed by enzymes. He ate broiled liver coated with sesame seeds and wheat germ for a powerhouse of such nutrients. He ate fresh fruits and vegetables daily. He also ate soybeans for good vegetable protein. For a good booster, he made this *Enzyme-Mineral Tonic:* To a glass of tomato juice, add two tablespoons of brewer's yeast, a teaspoon of desiccated liver powder, and a sprinkle of wheat germ. Stir vigorously. Drink as a healthy Enzyme-Mineral Tonic. These all-natural foods are powerhouses of minerals and proteins that are needed by enzymes to bring about better assimilation of other nutrients.

**Hormones Make Him a Happy Husband—and a Happy Wife is Grateful.** Once natural enzymes promoted a "magic rejuvenation" to his sex glands, Andrew C. was happy, healthy and like a "honeymooner" again. His happy wife felt grateful for Nature's powerhouse of youth—enzymes.

As you can see, hormones are present in minute quantities throughout the body glands. They remain inactive, inert and lifeless unless enzymes are present to whip them up and create new energy health and youth.

## *Five Ways to Use Enzymes to Rejuvenate Your Glands*

To give your glands that "feel young" feeling, give them enzymes. Here are five ways to use enzymes—using ordinary, everyday foods for a treasure of gland-rejuvenating benefits:

1. Use whole meat products that are sprinkled with raw seeds before using. Avoid processed meats which have nitrates and nitrites that destroy enzymes in the system. Potassium nitrate, a chemical preservative, is similar to saltpeter, the substance that was at one time believed to be given to prisoners to stifle their sexual urges. Avoid all chemicalized meats for better enzyme-gland health.

2. Boost your Vitamin C content (enzymes send this vitamin via hormones into the sex glands) with rose hips. These are the orange or red berries left after the rose petals have fallen. They are available in health stores and herbal pharmacies. Use as a natural sweetener. Sprinkle on whole grain cereals and over whole grain toast.

3. Boost your Vitamin A content (needed for nourishment of the body's glands and a collaborator of enzymes) by eating organ meats, halibut liver oil, cod liver oil, and liver in any wholesome form (except fried).

4. Your enzymes need the B complex vitamins to stimulate functions of the glands. These vitamins act as "energizers." Give them to your enzymes by eating lots of whole grain foods as well as brewer's yeast and wheat germ flakes.

5. Honey contains *aspartic acid.* This is a unique amino acid that is taken by body enzymes and used to feed your glands. It helps alert the pituitary and adrenal glands and also helps stimulate a sluggish thyroid. Enzymes need *aspartic acid* to create this hormone flow. Natural honey, sold everywhere, is a prime source of this amino acid.

So we see that your glands are the gateways to youth and health. Enzymes stand at the gateway. When they enter, they release the flow of youth-building hormones that perpetuate life from within and from without. Enzymes are available from natural foods. Give your body such healthful foods. Open up your body's gateways and release the supply of "youth rivers" that are enzyme treasures of a feeling of eternal youth and health with the sparkle of rejuvenated sex activity.

*Important Points:*

1. Enzymes are generators that stimulate the glands to issue forth their daily quota of youth-building hormones. Without enzymes, hormones remain inert and lifeless.

2. Enzymes in Vitamin E and cold-pressed oils act as natural propellants to prompt the pituitary to send forth its youthful hormones.

3. Enzymes in brewer's yeast and fresh fruit juice alert the adrenal glands to secrete needed "emergency hormones" to meet the challenges of stress and strain. These enzymes give you hormones to soothe the nervous system too.

4. Kelp and brown rice create enzymes and vitamins that send nutrients to the thyroid so it can provide you with a steady metabolism and a youthful appearance.

5. Yogurt enzymes are fermented substances that use minerals to nourish your parathyroids. This helps act as a "natural tranquilizer" and to give you a feeling of peace of body and peace of mind.

6. Guard against diabetes with a healthy enzyme-nourished pancreas, made possible by ordinary unflavored gelatin and lots of fresh fruits and vegetables. Avoid processed sweets and starches.

7. Raw fruit enzymes promote thymus hormones that issue white blood cells to guard against allergies, infections and viruses.

8. Male and female sex glands are supercharged by enzymes in outlined foods. Andrew C. became a lover instead of a loser with a five-step enzyme program that was easy to follow and had ingredients that were available anywhere.

9. Rejuvenate your glands with the outlined steps to enzyme enrichment using all-natural foods and reclaim any "fading" sexual capacity.

# 13

## *How Enzymes Create Youthful Energy and "Instant" Vitality*

Natural fruit enzymes have a "built-in" capacity to convert sugar into sources of youthful energy and "instant" vitality. These protein-like substances that are abundant in fresh raw fruit and their freshly prepared juices possess the power to transform natural sugar into a form that is speedily assimilated by your body. The benefit is that within a short time, the natural fruit enzymes are able to send a stream of "wake up and live" vitality shooting through your bloodstream, giving "instant" energy to your body and your mind too.

### *How to Use Fruit Enzymes to Wash Away Fatigue*

The key to vitality is in the proper use of fruit enzymes. These natural miracle workers have the ability to correct or prevent fatigue-causing problems caused by *hypoglycemia* or low blood sugar. Here is how this fatigue problem is caused–and how you can use all-natural fruit enzymes to wash away the symptoms and give yourself (body and mind) a healthy supply of "go power":

*Hypoglycemia Problem:* The pancreas is forced to secrete an excess of insulin to burn up the excessive amount of refined sugar (from processed and unnatural foods) that has been eaten. Because sugar is totally void of any nutritive value, it uses up a great many enzyme functions for assimilation. It also uses up much of the pancreas' supply of needed insulin. This hyperinsulin condition causes the sugar, or *glucose,* as it is called, which the enzymes have metabolized, to be burned up too rapidly. This leads to the problem known as *low blood sugar* in which the body and brain cells become starved for nutrition. Enzymes cannot gather enough glucose to send to the muscle and brain cells since the supply has been so rapidly metabolized. As a consequence, body functions suffer. The brain cannot function youthfully. Symptoms may range from nervous unrest to severe schizophrenic disorder.

**Using Fruit Enzymes to Provide Fatigue-Washing Vitality.** Enzymes in fresh raw fruit and fresh raw juices will take the body's supply of amino acids and use these protein substances to create a slow-and-steady metabolism of fructose (the natural energy-boosting sugar found in raw fruits) to give you a regular supply of youthful vitality.

Fruit enzymes are able to maintain a steady blood sugar metabolic rate that supplies a regular amount of energy to body cells from head to toe, including those in and around the brain. Fruit enzymes create a stabilized blood sugar process. These enzymes feed fructose to the body glands, especially the adrenal cortex. This gland uses enzymes in order to secrete its hormones that help correct the energy-producing blood sugar level, putting it in balance. Enzymes use adrenal cortex hormones to send slow-metabolizing but speedily energizing fructose throughout the bloodstream. This provides the natural and long-sustaining "lift" that gives you pep and vigor in body and mind.

**Fruit Enzymes vs. False Sugar Energy.** Enzymes use natural fruit sugars (fructose) for a slow and steady metabolism that will provide you an even level amount of youthful vitality and pep.

Conversely, refined white sugar, whether from a shaker or in processed and bleached foods, is devoid of enzymes and devoid of life itself. White sugary foods give a "false" energy. Refined sugar or starch (which is quickly changed into sugar) forces the body's enzymes to send a burst of energy to body and brain cells. The blood sugar level rises abnormally high and then drops abnormally low. The body's enzymes work frantically, stimulating the pancreas to shoot out a lot of insulin to metabolize the white sugar. This creates a seesaw or up-and-down insulin level that contributes to erratic and disturbed body and mind temperament.

For more lasting energy, your enzymes need natural fruit sugars to give your body cells a steady supply of energy that does not cause a vicious drain upon the pancreas and the supply of insulin.

### *How a Simple Breakfast Drink Provides Instant Youthful Energy*

Always busy Lisa P. relied upon the so-called convenience foods. She had little time to prepare a good breakfast for herself because she was constantly on the go with local garden clubs and women's organizations, and as an illustrator for the county newspaper. It was one deadline after another. Mornings meant a sugary breakfast drink that consisted solely of artificial ingredients and imitation flavors. With it was a sugary sweet jelly roll or pastry that was made from bleached, refined and devitalized ingredients. Lisa P. claimed she had no time to make an adequate breakfast and gulped down these "chemical foods" and then raced out of the house to her car to be off to her busy schedule.

This copious sugar intake was totally devoid of nutritive value. It sent her glands into an upheaval as enzymes from her system went into a panic to force her pancreas into a flood of insulin to transform the sugar into glucose. This internal churning gave Lisa P. a burst of energy as she was behind the wheel in the morning. But the sugar was burned so fast that within an hour she felt a letdown. She felt headachy and saw blurred objects instead of having clear vision. When Lisa P. experienced blackouts while behind the wheel, she decided to do something to save her health—and her life too!

**The Ten-Cent Breakfast Drink with a Million-Dollar Pep Power.** A guest speaker at one of the clubs to which she belonged told of preparing a simple breakfast drink—with just three ingredients that are available at every supermarket and are found in almost every pantry—and it took a few moments. The speaker said it cost her an average of only ten cents—but it gave her million-dollar pep power.

This miracle breakfast drink consisted of:

1. *Orange Juice.* A treasure of fruit enzymes that will take the natural *fructose* in the juice and use it to boost healthy energy. It is the natural sugar in fructose that creates this dynamic pep.

2. *Honey.* This natural sweetener contains predigested energy-boosting sugar. The bee has performed this predigestion, which means it goes to work immediately in your

system. Honey contains pure fructose and pure glucose. The secret here is that enzymes in the orange juice now take hold of these natural predigested sugars and send them immediately through the stomach wall, right into the bloodstream, *unchanged.* This gives you speedy energy.

3. *Unflavored Gelatin Powder.* Here is a prime source of protein. Enzymes use this protein to provide a steady metabolism that gives you a regular supply of vitality and energy. The enzymes use gelatin protein to spare the pancreas from a "shock" of insulin that predisposes you to low blood sugar and fatigue.

*Benefits:* When Lisa P. later made this enzyme breakfast drink by mixing one cup of orange juice with one tablespoon of honey and one envelope or one tablespoon of unflavored gelatin powder, she was preparing a potent enzyme tonic. In combination, the enzymes in the orange juice seized upon the predigested sugar of honey, then used the amino acids of gelatin to provide a steady burning in the system. This gave Lisa P. a gradual supply of steady energy.

Lisa P. later added a few slices of whole grain bread with a pat of butter. This gave her vigor-plus, thanks to the enzymes that worked in harmony to create energy that lasted until noontime. No more letdown. No more headaches. No more distorted vision. No more dangerous blackouts behind the wheel.

Lisa P. was grateful for discovering these three simple all-natural items, available anywhere and costing so little. But this simple ten-cent breakfast drink did give her million-dollar pep power!

### *The Simple Enzyme Food Program That Corrected Emotional Fatigue*

Some 300 persons who displayed symptoms of brain fatigue, premature senility and complete physical exhaustion were treated with a simple enzyme food program based upon wholesome, natural foods. It is reported[1] that these persons had such emotional unrest they had to be treated with sedatives, tran-

1. Harry J. Salzer, M.D., "Hypoglycemia," *Journal of the National Medical Association (1:66)*

quilizers and even electro-shock therapy. Little improvement was noted. Then doctors prepared a special enzyme food program that emphasized natural raw foods. The persons were not given anything artificial and were not given refined white sugar or products made from this devitalized "enzyme-dead" non-food. The three-step simple enzyme program:

1. Vegetables were fresh broccoli, Brussels sprouts, cabbage, carrots, cauliflower and others that were low in carbohydrates. As many raw (enzyme-rich) vegetables as possible.

2. All fruits were natural and unsweetened. The enzyme-rich fruits included apples, berries, and grapefruit.

3. *Taboo* were sugar-rich soft drinks and caffeine-containing coffee which would affect and disturb the enzyme sugar metabolism.

*Results:* Eight out of every ten of the persons were easily improved or recovered. A wholesome, natural, low-sugar, low-starch program had sent a steady supply of plant sugar into the system that could be utilized by the enzymes available in the foods and in the system. This helped the enzymes send a steady supply of oxygen-bearing blood to all body parts, especially to the brain cells. Enzymes used these natural plant sugars to invigorate the sluggish brain cells and promote clear thinking, youthful alertness and better health. Simple, yet youthfully effective, thanks to plant enzymes and avoidance of unnatural refined sugar products.

### *Enzyme Benefits of Plant Sugars*

Fresh fruits or their fresh juices contain enzymes as well as natural fruit sugars that work in harmony to create youthful mental and physical vitality. Fruit sugars are used by enzymes to partially replace fat or carbohydrate and to create a sparing effect upon protein. Plant sugars are used by enzymes to stimulate gastric secretion and motility, maintain better gland balance, normalize fat absorption, and create a form of youthful emotional health that is welcomed by all.

Enzymes use the fructose to boost a psychotropic reaction. A stream of natural sugar creates a youthful "electric stimulation" of the hypothalamus (lower segment of the brain) to help improve mental vigor. Enzymes need plant sugars to alert the adrenals to stimulate *epinephrine* and *norepinephrine,* the hormones that the body needs to meet the challenges and stresses of daily living. Enzymes use plant sugars to stimulate these "stress shields" and give emotional vigor to otherwise sluggish brain cells.

*Suggestion:* For healthful body-mind alertness, your daily food program should include lots of enzyme-rich fresh fruits and vegetables and their freshly prepared juices. For a speedy "Brain Awakener," be sure to drink fresh juices. It's the all-natural "pickup" that does more than give you a lift–it keeps you lifted in youthful vitality.

### *How to Use Enzymes at Breakfast for Day-Long Vitality*

Breakfast is the most important meal of the day. In the morning, your body metabolism is sluggish. Your enzymes are awaiting food which can then be used to send a stream of nutrients throughout your system to help give you "go power" with "staying vigor." Enzymes need vitamins, minerals and most important, *protein,* to send essential ingredients streaming through the interconnected network of neurons to the various glandular systems and to the brain itself. Enzymes must "feed" the brain with amino acids that they take out of protein, and help keep you mentally alert for much of the day.

For day-long vitality, you need a good high-protein breakfast that gives your enzymes the working materials from which they can make needed substances to nourish your brain, glands, organs and billions of body cells and tissues. There is no substitute for an enzyme-high breakfast for better and more youthful health.

### *Why Protein Is the Best "Brain Food" for Breakfast*

Of the three basic enzyme energy sources, protein appears to be the best "brain food" for breakfast. Researchers[2] released these findings:

2. *Annals of Internal Medicine,* Harvard University Medical School Report, Vol. XVII, pages 913.

1. *High-Carbohydrate Breakfast.* A glass of orange juice, bowl of corn flakes with sugar and milk, a slice of buttered toast and jelly. This furnished nine grams of protein. Enzymes created a swift blood sugar rise. It dropped quickly, reaching 69 milligrams two hours after breakfast. (Normally, fasting blood sugar is 90 milligrams per 100 cubic centimeters of blood.) It continued to drop still lower for another 1½ hours and did not reach the minimum 90 milligrams for five hours. The persons in this test experienced hunger, weakness and fatigue within two hours after breakfast.

2. *High-Fat Breakfast.* A breakfast consisting of a bowl of corn flakes with rich cream alerted enzymes to create a quick blood sugar rise. Enzymes used fat for a slower letdown. No hunger was experienced until more than four hours later. This breakfast had little effect on the rate of metabolism. Enzymes needed (but were denied) amino acids, and thus could not adequately nourish the body and mind. The persons eating this type of breakfast experienced sluggish mentality and physiological responses.

3. *High-Protein Breakfast.* A glass of skim milk, lean meat and cheese. Enzymes were now able to boost the sleepy rate of metabolism. The enzymes enabled these persons to remain at peak efficiency for six full hours. Enzymes maintained a healthfully elevated metabolic rate to produce steady energy of body and mind. The persons reportedly thought clearly and had a sense of youthful well-being. Enzymes raised blood sugar but used amino acids to keep the levels well above the minimum level. Enzymes needed the amino acids to give the persons a healthfully elevated level of youthful energy.

### *Protein Enzymes for Your Breakfast*

A high-protein enzyme breakfast should emphasize eggs, meat or fish, cheese, beans, peas or seeds and nuts. Any *one* of these high-protein foods can be used by your enzymes to send amino acids to billions of body cells (including those in your brain) to give you youthful vitality. *Note:* You will give yourself a double-enzyme thrust by eating a breakfast of raw seeds and nuts, together with a fresh fruit juice and a fruit salad. The seeds and nuts are prime sources of natural enzymes and plant proteins as well as vegetable fats. These give your body the working materials from which a steady blood sugar level can be maintained. To give your enzymes lots of vigor, be sure your breakfast has some protein to help keep you "emotionally nourished."

### *The Breakfast That Creates "Forever Young" Energy*

Jennifer K., a hairdresser who must work most of the morning in a standing or walking position, finds that this easy breakfast gives her so much vigor that she feels "forever young" in the beauty salon. Jennifer K. is able to be alert, youthful and vital from early morning until noontime (and even after that) because of this enzyme-rich breakfast:

A glass of milk; a bowl of fresh sliced fruit; one egg, and a slice of whole grain bread with a pat of honey.

*Benefits:* The raw fruit enzymes seize the protein in the other foods and use them together with the natural fat to create a slow but steady blood sugar metabolism. This is a balance that is harmonious. It feeds a steady supply of energy to her body and her mind. It is a natural breakfast, brimming with enzymes and nutrients that help create a "forever young" energy. Nothing complicated. All of these items are available at just about every food market. Many of them are probably in your pantry right now.

### *Seven Benefits of an Enzyme High-Protein Breakfast*

A balanced breakfast that contains fresh raw fruit, high protein, fresh raw juices, appreciable amounts of fat in the form of polyunsaturated oils or margarines will give your body the raw enzymes and raw working materials that can be used by enzymes to offer you these vitality-building benefits:

1. *Improved Reaction Power.* Reportedly, such a breakfast will give you better reaction power. You will be able to respond to stimuli with more youthful alertness.

2. *Better Increased Reactions.* Enzyme-created blood sugar harmony will help you face multiple-choice decisions with greater clarity and alertness.

3. *Neuromuscular Strength.* A high-enzyme, high-protein breakfast will help reduce or eliminate those physiological changes in the neuro-muscular system which might occur in the late morning hours after some stress. There is greater physical strength to meet the challenges and tensions of the day's tasks.

4. *Better Grip Strength.* There is more favorable grip strength and better muscular coordination and grip endurance. Enzymes use amino acids for this purpose.

5. *Stronger Endurance.* An enzyme high-protein breakfast helps increase physical endurance, giving better physical response.

6. *Improved Work Rate and Work Output.* Enzymes use the amino acids, together with natural plant sugars and vitamins to enable you to work at a better rate and to produce more favorable results.

7. *Improved Oxygen Supply.* Enzymes are able to send an increased stream of oxygen via the nourished bloodstream to feed the brain, improve thinking abilities and boost the power of the thinking apparatus.

Protein is needed by enzymes for the valuable amino acid supply that performs many health-building functions. When enzymes have protein to work with, especially during breakfast, you are able to enjoy youthful vitality, reduced muscular fatigue, and better emotional-physical responses. All of these benefits are available with healthful protein-high foods at breakfast time when your enzymes need working materials from which to build better health of your body and your mind.

### How to Build a Better Appetite for Breakfast

Many people are either in a rush in the morning or just cannot "work up" an appetite for breakfast. Here are some suggestions on helping to boost a better appetite for an enzyme breakfast:

1. To have a better morning appetite, eat less at the previous night's final meal. Try cutting out processed desserts. Eliminate those midnight snacks. They spoil your breakfast appetite.

2. Exercise before breakfast, if possible. Try a jog around the block, or even a few minutes' walk. Or do some early morning exercises in your home.

3. To put new taste into breakfast, select high vitamin-mineral-protein foods such as lean slices of meat, fish. Try a plate of crunchy raw seeds and nuts. These have a high enzyme supply and high protein too.

4. Eggs offer some six grams of essential protein (per egg) needed by body enzymes for the digested amino acid food. Try eggs soft cooked, poached or scrambled in poly-unsaturated oil. Throw in some protein-rich sesame seeds into scrambled eggs or an omelet. Or try hard-cooked eggs chopped with walnuts, green pepper or sunflower seeds. A chopped onion slice with some soybean oil tossed in helps make a breakfast feast brimming with enzymes and sure to perk up any "lazy" appetite.

5. Try this good raw "Protein Punch" that is instant protein for your enzymes. Add a cup of hot water to one tablespoon of brewer's yeast. Season with dried herbs. Sip as a morning beverage. Just one tablespoon of brewer's yeast gives you some 20 grams of protein, with a treasure of vitamins, minerals and enzymes too. The dried herbs also add flavor. In a few moments, body enzymes use the amino acid content for perking up your body and your mind.

6. Raw wheat germ gives you lots of good protein. About 24 grams in one-half cup. Mix it in a cup of milk or fresh fruit juice. Add a mashed banana. With a side dish of assorted nuts and seeds, you have a "dynamite" supply of vitamins, minerals, proteins and enzymes. In harmony, they give you enviable youth power via a perked-up metabolism.

7. Try natural unsweetened applesauce sprinkled with raw wheat germ. Or any fresh fruit together with sesame or sunflower seeds–nice and crunchy, and a powerhouse of protein and enzyme-creating energy. You might also try a

cup of yogurt with wheat germ, or healthful cottage cheese with sesame seeds or wheat germ. All add up to tasty excitement, together with the nutrients needed to stoke your body's fires to produce "instant" vitality.

### *Natural Enzymes for Youthful Health*

Enzymes in natural, raw foods have the energy principle or "vibration" needed to promote a boost of energy within the body. The enzymes alert your body's molecules and atoms, causing this reaction. The enzymes themselves do not become part of this process, but act as cosmic principles to promote this change. Enzymes produce the miracle process of *replenishment.* They create a feeling of youth and health. Feed your body with healthful, nourishing food that contains healthful enzymes, and you give your body and your mind the working materials out of which energy and vitality can be made.

#### *Important Points:*

1. Simple fruit enzymes are able to help balance blood sugar levels to correct hypoglycemia and to promote youthful vitality.

2. Lisa P. uses a simple breakfast drink to give herself day-long energy. She says it is a ten-cent drink with million-dollar pep power. Just three ingredients, available everywhere for a few pennies, give you a powerhouse of youthful vitality.

3. Some 300 "mentally sluggish" persons followed a prescribed three-step all-natural enzyme program and reportedly felt restoration of youthful health.

4. Protein is a top-notch "brain food" needed by enzymes at breakfast time.

5. Jennifer K. enjoys "forever young" energy with a healthy breakfast, brimming with enzymes and nutrients.

6. An enzyme high-protein breakfast gives you seven youth-creating benefits.

7. To boost appetite at breakfast time, try some of the seven suggestions. You can feast your way to better health and youthful vitality at the same time while you enzyme-rejuvenate your body and your mind with simple foods in easy-to-make combinations.

# 14

## How Raw Food Enzymes Create a "Feel Younger" Digestive System

Natural enzymes in raw foods are prime sources of magic-like ingredients that reportedly have the power to promote a feeling of digestive rejuvenation. Many European doctors are using raw foods as an all-natural way to help heal disorders of the internal organs. They have found that raw plant enzymes are able to help restore youthful vigor to the digestive system.

**The All-Natural Magic Enzyme in Raw Plant Foods.** A European doctor has discovered that raw plant foods (fruits and vegetables) contain a "magic enzyme" that is able to restore and regenerate the internal organs. Professor Werner Kollath of the Karolinska Hospital in Stockholm[1] calls these enzymes *auxones.* The word is taken from the prefix *auxo,* meaning "a form of combination" with a relationship to acceleration or growth. Auxones are either twins or members of the same magic enzyme family found in raw plant foods. Auxones are magic enzymes in that they have the all-natural ability to boost sluggish digestive functions and also propel other organs into a more youthful vigor and ability. Professor Kollath feels that these auxones or "magic enzymes" determine the growth of the plants. In the human body, they can perform the same internal regeneration and life sustenance. The doctor says that a deficiency of such magic enzymes can contribute to reduced health or "mesotrophy" in which the body's functions can undergo a decline. This may be

1. *Der Wendepunkt,* (4:49) Zurich, Switzerland.

similar to premature aging. The doctor feels that many people in the civilized world suffer from a state of "half health" because of a diet consisting of overprocessed and overcooked foods.

Professor Kollath explains that these magic enzymes or "auxones" in raw plant foods are sensitive to heat as well as to chemical preservatives.

*He shares the view of other doctors in that better health and digestive youth are achieved with raw, unprocessed fruits and vegetables as well as their freshly squeezed juices. These are the prime sources of the magic enzymes or "auxones" needed for digestive regeneration.*

He suggests that folks eat raw, nonprocessed plant foods and drink their juices as frequently as possible to help maintain a healthy digestive system.

**Raw Vegetable Enzymes Boost Mouth Digestion.** A Finnish researcher has discovered a powerful youth-building function of raw vegetable enzymes.[2] Professor A.I. Virtanen reports that *chewing raw vegetables* releases a shower of the plant's "magic enzymes" which then come into contact with their respective substrates or zymolytes (substances acted upon that need to be digested). The raw vegetable enzymes are then able to break up food particles into many smaller molecules so that they can pass more readily into and through the digestive tract.

*Benefit:* If you begin your meal with a large raw vegetable salad and then *chew thoroughly,* you will release the supply of "auxones" or magic enzymes right in your mouth! The benefit here is that food eaten will be predigested to a large extent in your mouth. This spares the vigorous effort needed in your digestive system. It is this simple program that can help make more food more readily digestible.

**Mouth Temperature Boosts Digestive Function.** Raw plant enzymes perform this "mouth digestion" very speedily because of the temperature in the mouth itself. This sends forth a supply of mouth or salivary enzymes that further "attack" food to be eaten so they can be more comfortably received and more healthfully handled by the digestive system later on.

**Raw Plant Enzymes Continue Digestive Work.** Professor A.I. Virtanen shares the opinion of other researchers that raw plant

---

2. *Report on Primary Plant Substances,* Helsinki, Finland: Biochemical Institute.

enzymes released through chewing are not completely destroyed after they have reached the digestive system. These plant enzymes or "auxones" are able to get into the intestine.

*Benefit:* The "auxones" help to detoxify the bacterial flora in the colon as well as normalize the bacterial environment. This reduces the quantity of decay-causing bacteria. This also creates youthful stimulation of the desirable lactic acid forming bacteria. All this is possible with this simple program: (1) Begin a meal with a large raw vegetable plate. (2) Chew thoroughly so that your salivary enzymes and mouth temperature can release the "auxones" from the plant food. (3) Chew the rest of your food thoroughly so that the digestive system is spared the task of complete churning. This helps share the "burden" of digestion and helps promote better assimilation of food. Simple, but very beneficial to your digestive system.

## *Why Raw Foods Help Create Youthful Digestion*

To enjoy better health, to guard against the ravages of degenerative illness, to enjoy a youthful power of resistance, you must give your body the raw working materials out of which "natural immunity" may be created. The youthful resistance to illness is often attributed to a more youthful digestive apparatus. While you may eat good food, you may not be *assimilating* all of the needed nutrients that build a natural fortress against degenerative illnesses. Youngsters have a strong enzyme system. In adulthood, there is a slight weakening or tapering off of this function. To help boost a sluggish system, raw foods contain the magic elements needed to create a rejuvenated power of assimilation—the key to natural immunity.

**How Raw Foods Protect You Against Aging.** A Swedish doctor reports[3] that *raw* and *nonprocessed* foods can heighten your powers of resistance to aging and also offer immunity to many degenerative diseases.

Dr. Henning Karstrom urges the eating of *raw plant foods* to help give your body the natural antibodies to fight off age-causing illnesses. He says, "It appears that the substances which form through enzymatic action in uncooked masticated foods are in

---

3. *Protectio Vitae,* (2:72) Stockholm, Sweden.

general of great health-giving importance. And, even though we do not precisely know the chemical structure and the physiological impact of these exceedingly important nutritional factors, we do know with certainty that they exist."

**Balanced Diet Still Requires Raw Plant Enzymes.** Dr. Karstrom emphasizes that even if you have a balanced diet containing every single one of the 50 known nutrients, and even if you eat a proper proportion of items from the basic four groups, *your health may still decline if you fail to eat raw and nonprocessed foods.* He explains that you may exist on such a diet for years and not feel ill. But neither will you feel entirely well. You will dwell in the age-causing area of *mesotrophy.* This is "the twilight zone of ill health" or "the incubation period." It is half-young, half-old, half-nourished. This may explain why so many persons succumb to degenerative ailments at a comparatively young age. This may also explain why many people are victims of premature aging of the body and the mind.

A balanced diet of healthful food is the first step toward youthful vitality. A program of raw and nonprocessed fruits and vegetables is the second and most decisive step toward better health and youth prolongation at any age! The magic enzymes in these raw plants boost mouth digestion and abdominal digestion—a double-action benefit of youthful assimilation.

### *How Raw Plant Magic Enzymes Cleanse the Colon*

The noted Swiss healer, Dr. Ralph Bircher, head of the famed healing institution and health spa, the Bircher-Benner Clinic in Zurich, also reports[4] that raw plant magic enzymes can help promote youthful health through a natural cleansing of the colon.

Dr. Bircher says, "While it was previously doubted that the enzymes reached the colon in an efficient state, it has now been shown that 60 to 80 percent of them arrive there unimpaired and—by oxygen fixation—establish anaerobic conditions in the intestinal tract, the medium in which the beneficial coli bacteria grow and multiply, and thus drive away the pathogenic ones."

At the famous Swiss Bircher-Benner Clinic, tens of thousands of persons have enjoyed digestive rejuvenation and "miracle" healing

4. Dr. Ralph Bircher, *Eating Your Way To Health,* New York: Penguin Books, Inc., 1972.

because *all meals begin with raw vegetables and sometimes chopped fruits.*

**Secret Rejuvenation Power of Magic Enzymes.** Raw plant foods contain special aromatic substances and enzymes which prevent digestive leucocytosis (a condition in which there is an excess of white blood corpuscles concentrated in the intestinal walls.) This condition impairs full assimilation. But raw plant foods contain magic enzymes and "auxones" which prevent this accumulation in the intestinal walls. This boosts digestive assimilation.

(Another secret benefit is that the excess concentration of white blood corpuscles leaves the blood deficient in these natural defenders and the body more susceptible to ailment—but if you eat raw plant foods, these needed white blood corpuscles will remain largely where they belong—in the bloodstream—so they can protect you against illness and aging.)

*Note:* Even if you eat cooked foods *after* a raw food, this leucocytosis will not occur. The secret here is to eat a sufficient amount of raw foods at the beginning of your meal so that your colon will have an adequate amount of protective enzymes to guard against the influx of these white blood corpuscles.

**Plant Enzymes Create Youthful Processes.** Dr. Ralph Bircher explains the secret of the chief benefit of enzymes in raw foods. There is an increase of micro-electric tension in the cell tissue of the whole body. Raw plant nutrition alone is capable of producing this.

*Benefit:* "With the increase of micro-electric tension, cell respiration improves, cell metabolism is stimulated, the powers of resistance and cell renewal increase, metabolism as a whole proceeds more economically and all natural healing processes are intensified. The most recent Japanese researches at the University of Kyushu have shown that raw plant foods prevent even hunger edema and hunger anemia, and that they strengthen the powers of resistance in the blood in its fight against disease-producing agents."

### *What Raw Plant Enzymes Can Do to Boost Digestive Health*

To look and feel young, give your body the needed amount of raw plant enzymes. Here is what these "auxones" can do to help boost digestive health and overall youth:

1. Raw plant enzymes will help promote the formation of red blood cells to enrich the bloodstream and build youthful health.

2. Raw plant enzymes will stimulate respiration and nitrogen metabolism of the cell tissues. This is an activity noted in youth.

3. Raw plant enzymes will improve the utilization of protein, taking out the needed amino acids to act as building blocks of the body from head to toe.

4. Raw plant enzymes help create a better mineral balance to control blood pressure.

5. Raw plant enzymes are able to send digestive nutrients to body glands to maintain a more favorable hormone balance. This helps promote better circulation, better healing of wounds, and improved pancreas and thyroid activity.

6. Raw plant enzymes are used by the digestive system to help restore and maintain a more favorable acid-alkaline balance.

7. Raw plant enzymes give the digestive system a helpful protein of extremely high biological value that promotes healing as well as building-rebuilding of the digestive tract and assorted organs.

In the words of the Swiss healer, Dr. Ralph Bircher, *"Begin each meal with raw food. Raw food before, not after, the cooked food. Whenever and however possible: No day without green leaves!"*

### *Why Unheated Oils Are Healthy for You*

In a series of reported tests,[5] a team of doctors headed by Dr. Rakel Kurkela have found that subjects who were given heated fats and heated oils were subject to weight increases. It is believed they were also shown to develop degenerative ailments and the ravages of ill health.

Dr. Kurkela reports that heated fats (oils, butter, and margarine) undergo unhealthy changes. Heating causes the development

5. *Zusammenfassender Vortag mit Literaturangaben,* No. 3 University of Helsinki, Finland, 1968, pp. 57-65.

of poisonous polymerization products. These destroy the enzymes and other nutrients. But more serious, heated fats can alter the body's functions and predispose to cellular destruction and the onset of aging.

Dr. Kurkela feels that *unheated oils are healthy* since they are prime sources of nutrients and the needed magic enzymes that enter into the regeneration of the digestive system and entire body as well. It was also reported that when oils are heated, their cholesterol-lowering benefits are greatly diminished. Since cholesterol buildup and atherosclerosis are byproducts of aging, it is reasonable to believe that years of eating foods prepared with heated fats may cause an elevation of cholesterol. This could be avoided by avoiding heated oils.

### *Dr. Kurkela's Five-Step Program*

Dr. Kurkela offers this five-step program that will give you many magic enzymes in oils as well as needed nutrients to create body harmony:

1. Do NOT fry foods in unsaturated oils such as safflower, sunflower, corn, soy, wheat germ oil, etc. These oils are prime sources of enzymes which take the essential fatty acids out of the liquid to be used by your body for better health. Heating destroys the enzymes. The essential fatty acids therefore may remain inert and lifeless.

2. When using oils, select the cold-pressed, polyunsaturated and nonprocessed variety. (These oils are sold at almost all health stores and many supermarkets too. The label should state "cold-pressed polyunsaturated.")

3. More and more margarines are being made with increased amounts of polyunsaturated oils. Therefore, *margarine should not be used for frying.* (Because fried foods of any kind are destructive to enzymes, these should be avoided.)

4. Butter is an animal fat and contains cholesterol. It is harmful when heated. Experiments show that when cholesterol is heated to high temperatures, it changes into a

substance called *methylcholantren.* This substance cau destruction, erosion and disorder of the body's cells a tissues. It is the forerunner to complete body deterioration. Avoid foods that contain heated butter.

5. Because olive oil is mono-saturated and has little influence on the cholesterol levels, Dr. Kurkela feels it may be suitable for heating. But you should avoid high temperatures. Do NOT use any fat when it darkens. NO food, if it must be fried, should be eaten if it turns a dark color. Keep frying to a bare minimum. Use olive oil for this occasion.

**How a Sales Demonstrator Corrected Colitis Upsets with Raw Food Enzymes.** Theresa E. was troubled with colitis as well as gaseous upsets. As a sales demonstrator, she felt embarrassed over her digestive upheaval. Standing before a group of customers, she found it difficult to concentrate on her work because her stomach felt "full," her innards were "bloated" and she had a "sour taste" that extended right up her throat to her mouth and tongue. Theresa E. lost much working time because of problems of "irregularity" that kept her confined to her home and convenient bathroom. It was during a chance moment when she overheard a radio lecture on the values of raw foods for correcting and healing digestive misbehavior that she went on the all-natural program.

Here is how Theresa E. rebuilt the digestive youth-power of her system by following a simple yet rejuvenating raw food enzyme program:

1. She began every meal with either a raw fruit salad or a raw vegetable salad. This included fruit before breakfast, vegetable before the noon meal and a large raw vegetable salad before the dinner meal. *Benefit:* Thorough chewing released the enzymes from the raw plant foods so they could help prepare the digestive system for further work.

2. Once a week, she would have a grated vegetable and sprout meal. She would flavor it with a sprinkle of safflower oil mixed with some lemon juice. This would be her main meal of the day. *Benefit:* The plant enzymes converged with the enzymes of the oil so that she received a full complement

of needed auxones. These stimulated the sluggish digestive enzymes so they could be ready for heavier work later on.

3. Daily, she would drink several glasses of fresh raw fruit and vegetable juices. *Benefit:* Plant enzymes were released through the juicing process so they could be speedily utilized by the sluggish digestive system. When juices are extracted from raw plants, they contain natural water of the finest quality, swimming with enzymes. Juices contain nourishing atoms and molecules in abundant volume. These enzyme-powered molecules nourish the digestive organs. Furthermore, juicing causes the plant fibers to be properly *triturated.* This process rips open the tough fibers of the plants to release their life's juices, or enzymes, so they can be used by your system for digestive renewal and regeneration. Just as the plants' life juices give them life, so they can give youthful life to your digestive system and the billions of body cells and tissues.

4. Theresa E. adjusted her cooking tastes and eliminated any and all fried foods. She would either bake, broil or stew foods. She would use polyunsaturated cold-pressed oils *unheated* for her salad dressings. This helped her enzymes use the essential fatty acids needed for digestive and all-over body rebuilding.

5. Theresa E. then decided to eat wholesome, fresh foods. She eliminated processed, packaged, dehydrated, pre-cooked or otherwise "dead" foods. With a simple change she soon released a stream of youth-building enzymes that worked avidly to build and rebuild her tired digestive system.

**Experiences Rejuvenation.** It took close to two weeks of this "back to Nature" enzyme program before she experienced relief. Now that her colon was cleansed, now that the auxones or plant enzymes helped to detoxify the bacterial flora and stabilize a healthful bacterial ratio, her intestinal canal and the adjoining organs felt youthful, healthy. Theresa E. no longer had stomach pains, no longer felt churning spasms, no longer was troubled with embarrassing "irregularity," no longer felt a "lump" in her stomach. Now she could face an audience with comfort and confidence. She looked the picture of glowing health, thanks to the all-natural program of raw food enzymes in plant foods available just about everywhere!

### *13 Health Benefits of Raw Food Enzymes*

When you follow the raw food way to digestive rebuilding, you reap the 13 health rewards offered by enzyme actions.

1. Raw plant foods require more chewing to release needed enzymes that take the digestive burden off the abdominal organs. Chewing also offers the teeth and jaws more exercise.

2. Chewing plant foods will bring about a strengthening of the mechanisms that cause salivary enzymes to come forth. This helps to "mouth digest" many foods and make them more easily digested.

3. Raw plant foods (being neither hot nor cold) will help preserve the teeth, tongue, mouth and digestive tissues from injuries often caused by very hot or cold foods.

4. Raw plant foods contain a natural proportion of nutritious matter that is needed for fibrous bulk. This nature-balance makes raw plant foods easily adapted by the body.

5. Raw plant foods offer a natural way to prevent a *balance* of harmonious flavors. They also guard against inharmonious combinations.

6. Raw plant foods are more healthfully assimilated by even the weakest digestive apparatus.

7. Raw plant foods contain more than just enzymes: they have a supply of vitamins, minerals, proteins, fats, carbohydrates and water in the organic and unimpaired state in which Nature created them. Such nutrients are as pure as possible.

8. Raw plant foods require chewing. This action exerts a physio-reflex response that calls forth sluggish digestive enzymes for their actions to follow.

9. Raw plant foods offer a satiety value since the taste buds are satisfied by chewing and there is an automatic control of overeating.

10. Raw plant foods (if organic) are usually free from the chemical preservatives that are found in canned, packaged and processed foods. Even if raw plant foods are not organic, they are natural since they have not been factory processed. They are preferable to packaged foods.

11. Raw plant foods do not spoil or decompose as rapidly as processed foods or as cooked foods.

12. Raw plant foods, if spoiled, cannot be "camouflaged" and passed off as good foods, as in the case of cooked foods. You are assured of good raw enzyme-rich plant foods when you look at the whole plant and see that it is fresh and unspoiled. This is not true with lifeless or cooked foods that can be disguised in appearance and taste (but not to your sensitive digestive system!).

13. Raw plant foods are truly all-natural "convenience foods." No cooking–no intricate recipes–just wash under cold running water, peel if necessary, then cut up and serve promptly. Raw plant foods are Nature's own convenience foods–brimming with natural goodness, natural enzymes, and natural health.

To enjoy the benefits of a "feel young" digestive system, feed your body raw food enzymes. It's the natural way to better health, beginning with your digestive system, the "heart" of perpetual youth!

### *Most Important Points in This Chapter Summarized*

1. Europeans discover a "magic enzyme" in raw plant foods that rejuvenate the digestive system and reportedly hold the key to youth.

2. Chewing raw vegetables boosts mouth enzyme flow and better stomach digestion and assimilation.

3. A Swedish doctor reports that magic enzymes in raw foods can forestall aging and offer immunity to many degenerative diseases.

4. A Swiss healer is able to cleanse the colons of ailing persons and restore them to youthful digestive and body health with raw plant magic enzymes.

5. Raw plant enzymes offer you seven unique digestive health benefits.

6. A Finnish doctor explains why unheated oils are prime sources of nutrient-carrying enzymes needed for digestive rejuvenation. He offers a simple, easy-to-follow five-step program you can follow in your own home.

7. Theresa E., a sales demonstrator, corrected colitis and related digestive distress with a simple but miraculously effective raw enzyme program.

8. Your body receives a minimum of 13 health benefits of raw food enzymes as listed.

# 15

## *How to Rebuild and Energize Your Organs with Natural Enzymes*

Raw food enzymes have the ability to transport nutrients to certain body organs and help promote strengthening, invigoration and a feeling of rebuilding.

More than 700 individual enzymes are known to be involved in the body's biochemistry, triggering reactions that furnish energy and rejuvenation for the heart, the liver, the kidneys, the nervous system, the organs of elimination and just about every other body part. Enzymes are able to bring nutrients to "starved" organs and thereby help to nourish them and keep them functioning at a youthful level.

The greatest benefit is that food enzymes are all-natural. You can reap the best of what Nature has to offer to enjoy the promise of perpetual youth through internal rejuvenation—thanks to food enzymes.

### *How Enzymes Use Seed Oils for "Cleansing" the Heart*

Your fist-sized heart weighs less than one pound. Every day it pumps some 5,500 quarts of blood over many miles of your circulatory system. In an average lifetime, your heart beats two and a half billion times (resting only between beats) and moves some 75 million gallons of blood with its life-sustaining oxygen and nutrients. Your heart is a powerful muscle and it needs to be treated well in order to serve you well.

**Polyunsaturated Seed Oils.** These are oils of fruit-vegetable-seed origin. These include corn oil, safflower oil, sunflower oil, soybean oil, walnut oil, sesame oil, and special salad dressings and margarines that are labelled as being largely polyunsaturated. Within these oils are enzymes and also a set of three essential fatty acids. These essential fatty acids (unofficially but adequately labelled Vitamin F) are used by enzymes for cleansing the heart, keeping it reasonably free of those hard age-causing plaques of cholesterol and problems of artherosclerosis or hardening of the arteries. To keep your heart clean, use polyunsaturated seed oils. Here is what they do to help keep your heart healthy:

**Enzymes Use Vitamin F.** Enzymes from the cold-pressed and nonhydrogenated oils work with digestive enzymes to remove Vitamin F from the oils. Enzymes need these three essential fatty acids found solely in cold-pressed and unrefined oils:

1. *Arachidonic Acid.* Enzymes use this oil-based nutrient to control cholesterol, to salt down accumulated deposits from arterial walls throughout the body, and to increase the body's self-protective flow of lecithin to meet the challenges of cholesterol. Enzymes must have this essential fatty acid found in cold-pressed oils.
2. *Linoleic Acid.* Enzymes use this essential fatty acid to build a better skin, to aid in metabolic functions, to shield your heart and your body against excessive radiation, and to help protect your heart against water imbalance.
3. *Linolenic Acid.* Enzymes use this essential fatty acid to help manufacture needed B complex vitamins, to protect against the buildup of cholesterol in and around the heart and the arteries. Your heart's health is often tipped into an imbalance if there is a deficiency of linolenic acid.

**Where Vitamin F is Found.** In cold-pressed and nonhydrogenated oils. This means that a hydraulic press extracts the oils from the seeds of plants. A mechanical expeller press squeezes out the oil from the seeds through use of a shaft revolving continuously against a fixed surface. Because this "expeller press" or "cold-pressed" method is used *without heat,* it produces oils that are brimming with the aforedescribed three essential fatty acids or Vitamin F, and a treasure of natural enzymes.

**How to Buy Such Oils.** Any large supermarket or health store will have a variety of different cold-pressed oils available. The label should state that there are NO added preservatives and NO hydrogenated fats. This gives you a large supply of Vitamin F and the needed cholesterol-washing enzymes.

**How an Electronics Engineer Used Seed Oil Enzymes to Ease Chest Pains and Boost the Health of a Clean Heart.** Benjamin A. was under severe tension. Excessive pressures of his job as an electronics engineer made him feel run down. He felt much older than his 35 years. He felt chest pains and short breath. He had a pale complexion. Often, he had to return from a field survey because he could not catch his breath; also, he felt a stiffening of his fingers and limbs, suggestive of arteriosclerosis. All this at an age when he should have been at the prime of youth!

*Easy Adjustments:* To begin the road back to coveted youth, Benjamin A. did more than use cold-pressed oils in his daily eating program. He *eliminated* hard or saturated and "dead" fats from his diet and *substituted* enzyme-alive Vitamin F-containing plant oils. This gave his body a stream of cholesterol-washing enzymes that helped scrub away internal debris. Furthermore, he was now able to feel more alert, and his complexion improved because of better oxygenation via the heart chambers. His fingers and limbs were flexible. Enzymes had used Vitamin F to help strengthen and improve Benjamin's internal organs. Now he is able to enjoy his youth with vigor and pleasure.

*More Adjustments:* To help give enzymes "room" to work, he made some more healthy adjustments. He followed this simple program:

1. Low-fat and low-cholesterol dairy products such as skim milk, buttermilk, cottage cheese, yogurt and low-fat cheeses.

2. Egg yolks are high in cholesterol. He ate just one egg yolk a week. Egg whites are cholesterol free, so he scrambled them, added a bit of oil, some herbal seasoning and a bit of cheese to give the taste of eggs—but without its cholesterol buildup.

3. Lean meats with all visible fats trimmed off.

4. Lean poultry (especially white chicken meat). He avoided poultry skin because the fat was concentrated beneath it.

5. Fish and seafoods (except shellfish) were good because they were high in protein, low in fat and high in those polyunsaturated fats needed by body enzymes for cholesterol washing.

6. All fresh fruits and vegetables and their juices. (The only exception was the coconut and coconut oil. This is highly saturated.)

7. He avoided any commercial pastries and baked products as they were made with highly saturated shortenings. For dessert, he had fresh fruit with a bit of honey.

8. He avoided processed breads, cereals and pastas. Instead, he used whole wheat and rye bread, plain rolls, low-fat crackers, whole grain and unbleached cereals (skim milk for a breakfast dish that is low in fat but high in enzymes).

9. For all spreads and oils, he selected those that were high in polyunsaturates.

*Results:* He felt that he could "bounce back" with youth, thanks to giving his body the enzymes and nutrients needed to protect his heart, his health—and his life.

### *How Enzymes Put New Life into Your Liver*

The liver is the largest gland in the body. Located on the right side of your abdomen just below your waistline, it is connected to the digestive tract. Your liver needs enzymes to help store iron and fibrogen, two valuable elements of the blood. Enzymes also trigger the liver to change carbohydrates into glycogen, which it then stores and releases to the body as needed. Enzymes prompt the liver to break down fats so they may be burned or oxidized in the cells. Enzymes enable the liver to pass bile (the residue left from the breakdown of red blood cells) into the excretory system.

Enzymes help store Vitamin D in the liver to be a source of strengthening the skeletal system. Without enzymes your liver would lie inert and useless and these functions could not be carried out.

**How a Floral Designer Used Enzymes to Rejuvenate Her Liver.** Selma M. looked faded and forlorn amongst the myriads of flowers in her shop. It was a sad contrast. She looked and felt older than her years. Her face was pale. Her energy was easily depleted. She always had to sit down after walking a little bit. The days grew longer and longer. In the midst of the bright floral displays that beamed youth and health, she looked depressingly faded and lifeless. Fortunately an organic flower enthusiast helped Selma M. reverse the aging process by rebuilding of the liver. The friend suggested a few enzyme programs that Selma M. followed. She was skeptical, but she had nothing to lose, so she agreed. The programs were:

1. *Liver Enzyme Booster.* To a glass of fresh vegetable juice, add one teaspoon of brewer's yeast powder, a rich source of Vitamin $B_{12}$. Sprinkle with herbs. Now add one teaspoon of desiccated liver powder, available at almost all health stores and herbal pharmacies. *Benefits:* Enzymes in the vegetable juice take the B complex and $B_{12}$ and Vitamin C from the herbs, together with vitamins from the desiccated liver, and use these to nourish the cells and tissues of the body's liver. It is the healthful and natural way to boost the function of a sluggish or "aging" liver.

2. *Whole Grain Foods.* An excessive amount of bleached white carbohydrate foods (Selma M. was always eating sugary pastries) can cause an overload of glycogen in the liver. This impairs its function. By switching to high-protein and high-vitamin whole grain foods, she spared the liver an excess of sugar. This enabled enzymes to perform their cleansing action upon the liver.

3. *Fruit Juice Enzymes.* Selma M. lived in a highly industrialized area. The problem here (as it occurs in many polluted areas) is that her liver became impaired by the inhalation of industrial poisons such as nitrobenzene, lead-containing gasoline, benzene and other hydro-carbons. Such

chemicals threatened to erode her liver. Selma M. was told to drink lots of fresh fruit juices. *Benefits:* Enzymes in the raw fruit juices will take Vitamin C and use it to help build and rebuild the tissues and cells of the liver as well as the entire body, offering a natural protective barrier against liver toxicity caused by air pollution.

4. *Lots of Raw Plant Foods.* Daily, Selma M. would eat as many raw fruit and vegetable dishes as possible. She would often devote an entire day to a raw food program. This gave her body a supercharging of needed enzymes to cleanse and regenerate the liver and body.

5. *Sprouted Foods.* A byproduct of pollution is the accumulation of ammonia within the body. To help change it to harmless urea, enzymes are needed. Sprouted foods are powerhouses of the B complex vitamins which are then taken by enzymes and used upon body buildup of ammonia. The enzymes use the B complex from sprouted foods to transform ammonia into a urea substance to be given off. If this ammonia is not washed out, it "attacks" the liver and creates internal toxemia.

*Selma M. Now Looks as Fresh as a Young Blossom.* A change to healthful foods, enzyme-containing raw foods and sprouts and better living soon helped Selma M. recover. Her organic flower enthusiast now said she looked as fresh as a young blossom, thanks to raw food enzymes that had given her a new life for her liver.

### *How Enzymes Help Keep Your Kidneys Doing Their Job for Youthful Health*

Your kidneys act as your body's filter plant, through which more than 150 quarts of liquid pass every day. Your kidneys are two large glands in the small of your back. They each consist of a great many tiny filters that strain waste products (chiefly metabolic byproducts) out of the blood, dissolve them in water and excrete them in the form of urine. Enzymes perform this function by prompting the blood to enter one end of each tiny tube in the kidney and then pushing it out through the smaller other end. Enzymes are the spark plugs that create this very effective filtering system.

An enzyme deficiency means that the waste products in the blood cannot be thrown off and toxic poisoning is now the threat.

**How Raw Fruit Enzymes are Kidney Cleansers.** Fresh raw fruits or their juices are especially helpful in soothing the kidney. Enzymes in the fruits create a healthful diuretic action so that the filtering mechanism runs efficiently. Enzymes take nutrients in the fruits and help metabolize them to mineral elements which promote a soothing washing effect.

Raw fruit enzymes reportedly also exert a significant bactericidal effect *in vitro* on infectious bacilli so that the kidneys can help remain immune to illness. Raw fruit enzymes possess a natural anti-infection benefit because of the complex *anthocyanin* pigments (natural disease fighters) and other polyhydroxy substances that create a natural inhibitory shield against bacteria. This protects the kidneys against the corrosive ravages of toxic assailants in the system.

**How a Raw Fresh Fruit Healed Kidneys.** Doctors report[1] that the banana, because of its low sodium content, its sparing action on protein, and its universal flavor appeal, is well adapted to a great many facets of therapeutic diets for the kidneys. It is recommended as a healthy food for the kidneys and the entire body.

**Helps Heal Nephritis.** This is a problem of kidney inflammation. Doctors treated several nephritis patients with bananas (substituted for the potato) to be eaten at breakfast, more bananas at noon and more bananas in the evening. Of course, other wholesome food was eaten together with the bananas. The doctors reported that the banana (probably because of its enzyme action) helped heal kidney inflammation, soothed gastrointestinal upset and was nonirritating and mildly laxative.

*Benefits:* Enzymes in the banana exert a sparing action on proteins, thus avoiding overloading of the kidneys. This helps keep down the blood-urea nitrogen level and lessens the severity of the syndrome. It is recommended that some nine bananas be taken daily, together with other supervised wholesome foods, for a period of three to four days. It is believed that this will help

1. Drs. M.G. Wohl and N.G. Schneeberg, *Clinical Nutrition,* New York: Paul B. Hoeber, Inc., 1950, p. 663.

renourish the body, soothe the kidneys and build better health. A miracle of kidney rejuvenation in a simple everyday fruit found just about in every food market for a few pennies apiece!

### *How Enzymes Can Help Soothe Your Nervous System*

A countless number of nerves transverse your entire body from top to bottom. Some nerves are involuntary in that they react without your conscious control. Many other nerves are voluntary and can be controlled. The foundations of your nervous system are your brain and spinal cord. From there, tinier nerves branch out to many parts of your body. Nerves are likened to the body's own electrical wiring system. Take good care of them. Improper care may cause a short circuit and nerve impulses can wreak havoc so that your body and mind become run down.

**How Enzymes Use Foods To Act As Natural Tranquilizers.** Enzymes take calcium out of such products as yogurt, skim milk, dairy products, and many green vegetables, and use them to soothe the nerves to help ease cramps in your arms and legs. Enzymes also take out minerals (especially magnesium) from the green coloring matter of fresh leafy vegetables and use them to soothe the network of your nerves. Enzymes also take out the B complex vitamins from whole grain foods (but not the bleached foods which are devoid of living elements) and use them to build the myelin sheath or covering of the nerve networks. Enzymes can do much to help ease tension by sending nutrients that act as *natural tranquilizers* for the nervous system.

**The Enzyme Tranquilizer That Relaxes an Accountant.** Because he was always under pressure, Christopher P. was the victim of "butterflies in his stomach," nervous disorders, temper flareups and fits of high-strung elation or emotional depression. He feared taking patent medicines because he knew they could be addictive. He had taken some barbiturates and suffered from dullness of memory and speech difficulties as well as vague hallucinations. He needed a *natural tranquilizer.* He was advised to take one made up of this mixture:

*Natural Tranquilizer:* In one cup of skim milk (or yogurt), mix one teaspoon of brewer's yeast flakes. Add some raw chopped sliced fruit and raw wheat germ. Stir and eat with a spoon.

*Benefits:* Enzymes in the fruit will take out the calcium from the milk and use it together with the B complex vitamins from the yeast and wheat germ to create a soothing effect upon the body.

*Results:* Christopher P. found that this Natural Tranquilizer gave him a partial relaxation benefit. Since he did not give up his high-strung way of living and his hectic schedule, he could not bounce back with youthful relaxation. But this Natural Tranquilizer, taken several times daily, did help insulate him against further corrosion of the nervous system.

### *Two Plant Juices That Provide a Natural Sedative Benefit*

Nancy O. had so many iatrogenic (drug reaction) responses from one medication after another that she feared for her life when considering chemotherapy. She was also troubled by financial reversals and the burden of trying to support her family and maintain her household at the same time. She needed a *natural sedative* that would soothe her tensions and make her feel better able to cope with the stresses and strains of her daily tasks.

**Mixes Two Vegetable Juices.** She was told to mix two simple but relaxing vegetable juices. In combination, they reportedly would exert a natural sedative action that would ease her tensions, soothe her frazzled nerves, make her feel naturally happy.

Nancy O. mixed equal portions of *celery juice* and *cucumber juice.* These were available at almost any health food store or in many large food markets. Sometimes Nancy O. would juice them herself: she would put equal amounts of freshly washed and scrubbed celery and cucumbers through her blender, make a puree and drink it slowly.

*Celery Enzymes:* Celery enzymes take out the vegetable's supply of magnesium and iron, both of which are known for soothing the nervous system. Together with celery's vitamins, the result is a powerhouse of nerve-soothing nutrients.

*Cucumber Enzymes:* Cucumber enzymes take out the high magnesium and the very high potassium content as well as many vitamins and other minerals. These two minerals named are very valuable in helping to soothe and relax the nervous system. The phrase "cool as a cucumber" is traced to the presence of these two minerals. They are taken by enzymes and propelled throughout the bloodstream to wash, coat and cleanse the nervous network and make it respond with sparkling strength to the stresses of daily living.

**Natural Sedative Works Health Miracle.** Nancy O. would drink as many as four glasses of this Natural Sedative daily. The enzymes immediately took the minerals and nutrients from these two vegetables, used them to restore the myelin sheath of the nerves, and relieve the harsh grating and counteract the degeneration caused by tension. Nancy O. found that her nerves were soothed and that her jittery fingers and nervous stammering subsided. She was no longer jumpy. She felt she could cope with tensions—thanks to the Natural Sedative that was brimming with enzymes needed to use minerals and other nutrients for the rebuilding of the shield covering of her nervous system. But most important to Nancy O. was that the tonic was all-natural!

Yes, self-renewal begins with regeneration of the internal body organs. Enzymes promote the action of nutrients to nourish, replenish and regenerate your vital organs. Enzymes help keep your body organs young. Result? You look and feel young too! You feel young all over! Raw food enzymes can make the youth dream come true.

*In Review:*

1. Raw food enzymes use substances in the foods to help promote healing and regeneration of body organs.

2. Enzymes use polyunsaturated seed oil substances (known as Vitamin F) to help cleanse the heart and guard against cholesterol-arteriosclerosis buildup, a forerunner to old age illness.

3. Benjamin A. used seed oil enzymes to ease problems of chest pains, short breath and premature aging. A simple yet youthfully effective remedy—all natural too.

4. Selma M. put new life into her liver with an easy and natural Liver Enzyme Booster tonic and a five-step program.

5. Raw fruit reportedly acts as a kidney cleanser and healer. Fresh raw fruits promote rejuvenation benefits to the body's kidneys.

6. Christopher P. uses a Natural Tranquilizer that eases tensions. Nancy O. uses a two-ingredient Natural Sedative that restores youthful vigor to her frazzled nervous system. Enzymes may hold the key to soothing the nervous system the natural way.

# 16

## *How Home Water Healers Help Enzymes Wash Pain Out of Aching Muscles*

The "pain-washing" benefits of water have been known for thousands of years. Even before mankind knew the intricate workings of his body, he was aware that bathing promoted a feeling of internal rejuvenation that helped melt away many aches and pains. Instinctively, he decided that if it made him feel youthfully good, then it was an acceptable means of healing. The ancient Greeks who worshiped youthful health in all its phases built temples near watering places, beside which they erected statues of Aesculapius, their god of medicine. They were firm believers in the healing power of simple water.

**Water, Nature's Healer.** The Romans also placed great faith in the natural healing power of water. Their physicians would prescribe regular bathing as a means of easing aches and helping to wash pain out of a worn body. For several thousands of years, throughout the world, almost all civilizations have looked to water as a means of restoring a feeling of inner youth and resilience.

The Orientals would soak in natural hot springs for hours at a time, as a means of melting away the stiff aches that afflicted their hard-working muscles and joints.

The Europeans have a tradition of water healing that has helped countless numbers from ancient days to our modern times. The famed Finnish, Russian and Turkish baths reportedly helped steam open clogged pores and thus invigorate the sluggish circulation which would then wash out impurities. Enzymes could now work in a "clean" and "washed" body environment to help send nutrients that promote healing of tired and aching muscles and joints. While knowledge of enzyme healing was unknown, the Europeans knew only that water bathing helped heal their painful spasms and knotted up muscles. That was enough for them.

The famed American hot springs have long been used by the early American Indians of Arizona and California. They too looked to water as a mystical or magical healing force. It is reported that many Indians would immerse themselves for hours in such "magic waters" and emerge with youthful bodies, vigorous muscles, with vanished aches and pains, ready to go forth in their arduous tasks and labors. To them and to modern people, the use of "hot springs" is known for refreshing-relaxing-rejuvenating a tired body and promoting relief of knots and painful spasms. Much is traced to a self-cleansing, as well as a form of "enzyme euphoria" which enables these magic substances to wash away the clogged debris that is the forerunner of muscle and joint tissue degeneration. Water has the healing effect that is known for invigorating sluggish enzymes and boosting their internal cleansing powers. Home water healers are easily made today in the privacy of your own bathroom for the benefit of better health through enzyme activation.

### *Three Home Water Healers for Enzyme Activation*

Using ordinary water from your bathtub and ingredients available at almost any herbal pharmacy or ordinary supermarket, you can make a home water healer in a few moments—for a few pennies—and soak yourself to freedom from nagging aches and pains. Here's how:

1. *Thalassotherapy for Taking the Ache Out of Backache.* The name is derived from the Greek word *thalassa* meaning the "sea" or "ocean." In ancient days, Greek physicians would prescribe bathing in sea water as a means of reviving sluggish body substances that would help correct "deposit accumulation" on body joints that contributed to backaches. Today we know that sea water bathing is an invigorating way to alert enzymes that start a self-cleansing activity to scrub or wash away the toxic crystals that inhibit free body movement. A backache is often the symptom of a

sludge-covered arterio-skeletal structure. A home sea water bath can help alert sluggish enzyme action that will then help promote body metabolism and wash away excess sludge.

**How Walter Z. Used a Home Sea Bath for Backache Relief.** Because his work as a construction man calls for much heavy lifting, Walter Z. is subject to recurring backaches. He finds that if he takes a Home Sea Bath for no more than 45 minutes, it makes him feel limber, resilient and ache-free. He turns ordinary tap water into ocean water by adding a cupful of kelp–a powdered seaweed product available at all health stores and herbal pharmacies and many food markets–to a tub of water that is comfortably warm. The minerals in the kelp seep through the open pores of Walter's body, nourish the skeletal structure, and propel enzymes so they can perform their internal cleansing benefit. He soaks himself for 45 minutes to one hour. The comfortable warmth of this ocean-type "thalassotherapy," used by the ancient Greek physicians as prescribed healing for pains, gives Walter Z. such freedom from backache that he is able to work like a youngster the next day. Thalassotherapy makes Walter Z. feel "young all over."

2. *Mud Baths Help Strengthen Aching Bones and Muscles.* This ancient healer is now featured in many beauty and health spas throughout the world. It calls for immersing yourself in a "bath" of oozing mud. It is believed that the treasure of minerals and enzymes found in the soil enters through the body pores and thereby promotes an alerting of the body's own enzyme structures. Once the body's enzymes are alerted, they are able to help correct problems of internal toxemia or debris which often inhibit free movement of joints and muscles. In effect, enzymes from the mud are propelled by the minerals found in the mud to perform a process known as "autolysis" or self-loosening within the body. This helps cleanse out internal debris and the enzymes can now work in a clean internal environment. After a mud bath, just wash off under free-flowing warm water and then comfortably cool water. You should emerge looking and feeling sparkling clean and resilient from the inside to the outside.

**Judith L. Uses Home Mud Baths to Put Youth into Her Bones and Muscles.** Judith L. obtains fuller's earth, a substance that can approximate a mud bath, from her herbal pharmacy. It costs very little. Just a handful in a warm tub and she has a healthful mud bath right at home. She soaks herself in the home mud bath for about an hour. The magic minerals and magic enzymes from the home mud bath are able to wash away the debris from the external-internal pores and actually slough away the toxic wastes from the system. Once this occurs, body enzymes are then invigorated and able to perform an internal cleansing action that helps wash out the toxic materials that have impeded the youthful resiliency of bones and muscles. Judith L. finds that just a once-a-week home mud bath is able to keep her free from the painful spasms of arthritis-like distress. Her fingers are more flexible. She has a tighter grip too. The mud bath is a simple and all-natural effective healer from the past, and a natural enzyme-alert way to better health.

3. *Aromatherapy Helps Soothe Muscle Spasms and Ease "Knots" in the Body.* The use of fragrances and aromas has been known for countless centuries. By inhaling fragrances from herbs or even from ordinary sea water, the body's responses become more alert, more activated. Just as oxygen is the breath of life, so will certain aromas promote internal rejuvenation. Aromas help revive sluggish body processes. The famous "smelling salts" of the days when women were fashionable when they fainted illustrate the self-revival powers of various aromas. Aromas from herbs or sea water have the ability to stimulate the sluggish body processes. Inhalation will invigorate sleepy enzymes and other body organisms so they can speed up their tasks in boosting body metabolism. Inhalants are able to create a beneficial physiological response. They also have the ability of being able to alert and activate enzymes to promote body cleansing which will enable muscles and tissues to enjoy resilient responses.

**How Dorothy D. Uses Aromatherapy to Take the "Knots" Out of Aching Muscles.** While Dorothy D. experienced some relief from nagging muscle aches by using hot pads and medicated ointments, she wanted a more comprehensive feeling of relief. So she would take a nightly aromatherapy bath right in her own tub. She would take a handful of any aromatic herbs and let them steep in the comfortably warm water of the tub. Then she would immerse herself, breathe deeply and feel herself "loosening up" in her body and mind. About one hour of such an aromatherapy bath each night would work miracles of relief. Her muscle spasms were soothed and the "knots" became "untied" so that she could discontinue using hot pads or medicated ointments. Soon Dorothy D. was able to twist and bend without the knife-like pain that had always shot through her body. She looked and felt as agile as a young gymnast, thanks to aromatherapy!

**How Home Healers Perform Their Magic Muscle Relaxation Benefits:** Sea water or herbs act as *diaphoretics.* That is, they help open up the body pores and then alert the enzymes to surge through the circulatory system and permit more perspiration and waste matter to escape. Once the enzymes are able to discard this internal clogging, they can then work to send nutrients to the cleansed muscles, cells, tissues, bone marrow, body organs and systems. These enzymes are then able to promote healing and regeneration. Muscles are usually the first to benefit in that they are cleansed, nourished and made more resilient. Now the body is remarkably liberated from nagging aches and pain symptoms. To promote this healthful healing, home healers prompt enzymes into this natural process and help give you youthful limbs and muscles.

**Open Pores, Cleanses Body, Refreshes and Relaxes.** Home healing baths help remove the dirt that clogs the hundreds of pores that line the surface of the skin. Enzymes are now able to work from within, when they are prompted by external stimuli. Body enzymes now promote cleansing of these "pipes" or skin pores. They loosen and help slough off the encrusted toxic wastes. Once the body is able to discard its accumulated waste matter, enzymes are able to work in nourishing the clean internal organs. This helps to refresh and relax and also regenerate muscles and sinews so that the entire body is able to bounce back with youthful agility.

*Suggestions:* Your bathroom should have several green growing plants. The benefit here is that they take in stale carbon dioxide and give off oxygen, thus giving your bathroom a pleasant environment that is similar to that of the great outdoors. Also, have a washcloth or a "loofah" (sponge or mitt made of vegetable fibers) for rubbing all over your body. This helps stimulate skin friction and rub away debris so that body enzymes can work in a clean environment. A vaporizer that sends forth a stream of fragrances is also soothing, promotes better inhalation, and alerts the sluggish olfactory (nasal) mechanism to accept this gentle stimuli. All of this will do much to help your body enzymes perform more efficiently. They will help soothe your muscles and make you feel rejuvenated.

**Roman Oil Bath.** Roman warriors reportedly took this luxurious yet amazingly simple bath to help rejuvenate their worn-out muscles. Now you can enjoy the vigor and strength of a Roman gladiator with this same bath in your own tub. Here's how:

In a tub of comfortable warm water, pour one cup of oil (the Romans preferred olive oil). Now add a teaspoon of your favorite perfume (the Romans preferred rose geranium oil, currently available at many pharmacies and some health stores). You may also add a tablespoon of any liquid detergent shampoo. A herbal shampoo is considered best. Let these ingredients mix with the warm water. Then just soak yourself in it.

With your palms, massage your tired muscles while you enjoy the soothing feeling of the Roman Oil Bath.

*Benefit:* The modern herbal shampoo is able to break the oil into millions of tiny globules that cling to your body pores. This does more than just cleanse–it helps make your skin completely "ventilated" and "aerated," thanks to the action of the herbs and the perfume. Now your body pores can breathe clearly. This sends a stream of life-giving oxygen to your enzymes. Thus alerted, they are able to send body nutrients to your body organs and muscles. This helps promote healing and nourishment of tissues and cells so that collagen, the cement-like substance, can bind up the fragments and soothe the damage so that muscles become regenerated from within.

This helps soothe the aches and pains of modern living. Small wonder that the physically violent Romans used this Roman Oil Bath as a "youth healer." They would emerge feeling refreshed, rejuvenated and ready for battle again. While you may not engage in gladiator battle, you will be better able to meet the challenges of our modern pace of fast living with a "young muscle" body. Do as the Romans did–enjoy a Roman Oil Bath for an hour a night! It makes life worth living.

### *Contrast Bath for "Wake Up and Live" Enzymes*

Anthony J. is the victim of recurring muscle pains and symptoms of arthritic "gnarling" of his limbs. He operates a one-man floor cleaning service and must operate heavy equipment and vacuum cleaners. Frequently Anthony J. was so doubled over with aches that when he bent to plug in a floor-washing device, he experienced a "jolt" in his back and felt stabbing pain shooting all over his body. It took him a long time to straighten up and be able to work again.

**Natural Method Promotes Better Relief.** Because Anthony J. has so many allergic reactions to chemical drugs, he was compelled to seek a natural way to relieve his body aches and recurring pains. One client told him that "good, old-fashioned contrast baths" work wonders. Anthony J. had nothing to lose (except his muscular pains) so he tried this program:

1. Soak in a tub of comfortably warm water. To promote better relaxation, put a small rubber pillow beneath the nape of your neck. Soak and scrub and do gentle exercises of just body motions while you are in the tub. Just soak up to 30 minutes.

2. Now let the warm water run out. While the water is running out, turn on the cold water full force. This gives you a *blending* of the warm water (slowly running out) and the new cold water. Mix both waters and splash all over yourself. Continue this mixing and splashing and self-massaging until the water gets cooler and cooler.

3. In five minutes, your entire body will glow with tingling vitality and youthful resiliency. You should feel refreshed all over.

4. Now rub yourself dry with a Turkish towel. Rub very briskly to help further tone up the micro-venous carrying enzymes located near the skin surface.

5. If possible, wrap yourself in a terry cloth robe and then just relax in bed while the body processes become self-renewed in their stimulated processes.

*Benefits:* Anthony J. used this simple program, together with the natural way to health through wholesome foods, and he found that he felt so "lubricated" that he could now bend over comfortably and no longer experience the painful muscle spasms as before. The benefits of this all-natural Contrast Bath include alerting a sluggish circulation system to boost the flow of somewhat sleepy enzymes. Once Anthony's body enzymes were able to function with vigor, carrying nutrients to his billions of body cells and tissues, they could then nourish his arterio-neural network and the cellular tissue respiration process to promote better muscular vigor. It was a welcome relief to Anthony. Now his body was resilient. His fingers had a better grip. He could work with the agility of a youngster—thanks to the stimulatory benefit of a simple and all-natural Contrast Bath. Just 40 minutes—for a "forever young" feeling of youth and vitality.

### *"Instant Energy" from the Quick Bath*

Enzymes can become supercharged with vitality by a simple *five-minute bath.* (This is actually a shower, but you will derive more soothing benefit by plugging the tub and letting it fill up while you stand beneath the shower.)

Turn on a shower of *comfortably* hot water and stand beneath it. Let the sharp tingly needle-like "fingers" of the shower work over your entire body. Stand beneath the *comfortably* hot shower for just three minutes.

Now gradually let the shower water become colder and colder. Let the cold water "fingers" alert your entire body from head to toe. After one minute, leave the tub and briskly rub yourself dry for just one more minute.

Altogether, you've let the hot and then cold water, with the towel rubbing, alert your body for just five minutes. The benefit here is that the sharp tingly "fingers" have whipped up your sluggish enzymes. Now they can bring nutrition to your billions of body cells and tissues. In particular, the awakened enzymes now send a store of *collagen,* the cement-like ingredient made from Vitamin C, to your damaged cells and tissues. Once they are rebuilt, your muscles, joints, limbs and just about every part of your body will respond with youthful vitality. All this and much, much more in the way of looking and feeling young, thanks to water-invigorating enzymes.

*Highlights:*

1. Water is an ancient healer to wash pain out of aching muscles.

2. To activate enzymes and help them promote healing of gnarled or "knotted" muscles, try any of these three home water healers. They reportedly soothed Walter's backache, rejuvenated Judith's muscles, soothed muscle spasms for Dorothy D. Simple yet pleasurably effective.

3. Enjoy an enzyme washing program in your own bathtub. Try a Roman Oil Bath for "bounce back" resiliency.

4. A simple contrast bath soothed Anthony J. and gave his aching back "wake up and live" flexibility.

5. Enjoy "instant energy" with a quick five-minute bath. Follow the simple and all-natural step-by-step plan for young muscles.

# 17

## *The "Enzyme-Slim" Program to Melt Pounds While You Eat*

Raw food enzymes have the ability to transform eaten food into a form that can be utilized by the body. They help metabolize starches, sugars and fats so that these become "burned" off and not stored as excessive fat. An abundance of enzymes means that your body is better able to convert ingested food into energy-producing vitamins, minerals, proteins and essential fatty acids. A deficiency of these enzymes means you have a slowed-up metabolism and food may turn to fat which contributes to unnecessary or excessive poundage. The key to remaining slim calls for a more healthful and enzyme-rich program. You should be able to enjoy many of your favorite foods and still lose, when you give your metabolism the needed enzymes that melt away pounds and help keep you slim and trim.

### *Five Ways to Give Yourself a Supply of Fat-Melting Enzymes*

Here is a simple five-step program that helps give your body a goodly supply of the needed fat-melting enzymes that help keep you youthfully slim:

1. The total amount of food eaten should contain a *higher* proportion of raw foods and a *lower* proportion of cooked foods. *Benefit:* The enzmyes in the raw foods will be used by your metabolism for the burning up of eaten foods and will thereby control the amount of fat remaining in your body.

2. Begin each and every meal with a raw fruit or a raw vegetable. A salad is best. *Benefit:* The raw fruit introduces a shower of fat-melting enzymes in your system to *prepare* in advance for the fat-containing food to be eaten. This helps boost enzyme-digestive action and lessen fat buildup. Another benefit is that the raw foods help give you good satiety value and there is less tendency for you to overeat.

3. Fat is essential to your health but the *right* type of fat is even more decisive in maintaining health. Select polyunsaturated and nonhydrogenated cold-processed oils as your fat. These are available in almost every supermarket and all health stores. *Benefit:* These fats contain three essential fatty acids needed by enzymes for the metabolism of ingested foods. They help control the appetite and also serve to replace the more "solid" fat accumulations from hard fats such as butter, margarines or animal fats.

4. Omit or reduce the use of any unnatural sugars or starches. *Benefit:* Substitute with fresh fruit sugars which give you speedy energy and less fat accumulation. Refined white sugar and bleached starch products are transformed into fat and obesity. Enzymes cannot metabolize them so speedily. Avoid these unnatural sweets and replace with moderate amounts of natural sweets (honey, maple syrup, and molasses) and lots of fresh fruits with natural fruit sugars that spark the function of enzymes to digest other foods.

5. Increase the amount of meatless protein to nourish your body and also to help guard against nitrogen overload which predisposes to obesity. *Benefit:* Enzymes transform protein into usable amino acids. A byproduct of this process is that of nitrogen which combines with water to create fat. To help control this water-logging, try to reduce your meat protein intake. Select good raw vegetables, soybeans, bean sprouts, whole grain cereals and whole grain bread products for meatless protein. This helps guard against the overload that is often the forerunner of obesity. It also helps give enzmyes a chance to metabolize other foods and not have to work avidly upon protein. This too creates better overall body metabolism and a slimmer you!

## 25 "Eat and Reduce" Ways to Better Enzyme Slimming

To create a better "enzyme environment," here are 25 ways to eat and still reduce, thanks to the miracle fat-melting powers of enzymes from your body and from raw foods.

1. Foods eaten should be fresh. Do *not* eat leftovers because you feel guilty about wasted foods. Let it go to waste—instead of to your waist!

2. To help activate your enzymes, eat a meal standing up! This helps boost calorie burning and also helps keep you slim. But be sure that you are comfortable while standing and not under any strain.

3. Start the day right with a whole grain, enzyme-rich breakfast that includes lots of fresh raw fruit. This gives you the needed enzyme intake to burn up foods eaten later on. It also helps control your appetite and ease the urge to eat more.

4. If you must snack, select raw vegetables that help give you enzymes and satisfy your appetite at the same time. You'll sleep better if you go to bed with a light stomach. If you just can't control eating at night then go to bed earlier!

5. Whatever you eat should have nutritive value. *Omit* those "non-foods" such as sugar or refined flour—any products made with these high-calorie but nonnutritive foods. Omit pastries, cakes, doughnuts and bleached macaroni or bread products. If it has no nutritional value, skip it!

6. Eat several small meals daily to help control your appetite and give your digestive system a steady, time-released supply of fat-burning enzymes. Do NOT skip any meal. The problem is that when you force yourself to miss a meal, a large supply of stored fat rushes into your bloodstream. This increases the fat supply in the blood and spells danger to your heart, your health and your weight too! Enzymes cannot metabolize an excess of this fat. So eat a few meals daily, satisfy your appetite, and give yourself needed fat-burning enzymes.

7. Eat until the sharp edge of your appetite is soothed. Do NOT eat to get full. Gradually, as you cut down on the excessive portions, you'll look and feel slimmer. Enzymes will be able to metabolize ingested foods at a comfortable rate.

8. If you're still "stubbornly fat," try this trick–reverse your meals. Eat dinner in the morning and breakfast in the evening. The benefit here is that enzymes can burn off calories eaten in the morning at a better rate than those eaten in the evening. It's the easy way to enjoy lots of foods with less calorie buildup.

9. Eating fat is helpful! Use unsaturated fats to help soothe your appetite and keep you slim. Enzymes are better able to metabolize liquid fats than hard fats. Just make a simple switch and omit hard fats and you'll help control fat buildup.

10. To slim down and keep slim, try the six small meals a day program. Frequent eating means smaller meals and a steadier blood sugar level. This also enables enzymes to better metabolize fats into usable substances. It helps your enzymes boost metabolism for speedier and more complete combustion (burning up) of food eaten.

11. Soybeans are wholesome and also help you keep satisfied on less fattening food. Enzymes take lecithin out of the soybeans and use this substance to control fat buildup and to spark a "cholesterol washing" rhythm that keeps your heart and body healthy. Enzymes will use lecithin to remove fat deposits from some body parts and put them where they belong, or else metabolize them right out of your body! These benefits are yours in soybeans, the "meat without a bone," as it is called in the Orient.

12. For beverages, drink fresh raw vegetable juices, herbal teas and coffee substitutes. Avoid coffee. This non-food is high in caffeine and contains noxious ingredients that punish the pancreas, forcing this gland to send forth excessive insulin and leading to a problem of erratic blood sugar. To soothe the appetite, drink raw plant juices which are brimming with enzymes that help satisfy the eating urge and promote better health too.

13. Avoid artifically sweetened foods. These contain chemical additives that destroy body enzymes and impair health. Furthermore, you need to readjust your taste buds for *natural* sweets in fresh fruits and vegetables, so you should avoid chemicalized artifically sweetened foods of any sort.

14. Physical activity helps boost sluggish circulatory responses and stimulate enzymatic burning of fat. Keep yourself active. With your doctor's approval, join any local supervised exercise group.

15. For a healthy enzyme-high and calorie-low lunch, try raw salads, or a cup of soup with salads. To boost the enzyme content of soup, sprinkle with raw wheat germ.

16. Here's a top-notch enzyme breakfast that gives you lots of "go power" and "enzyme power" to ease appetite and keep you slim. Soak several tablespoons of oatmeal overnight in some water. Next morning, add chopped nutmeats and fresh fruit slices. The benefit here is that the fruit contains *fructose* (natural fruit sugar) which is speedily taken up by the enzymes in the nutmeats, then propelled by the vitamins in the oatmeal to make you feel supercharged with energy and vitality. You don't feel hungry till well after noon.

17. Looking for an Enzyme Pep-Up Elixir? Just run a cup of unflavored yogurt, together with any raw vegetable, through a blender. Grate some nuts over it. Spoon in some lemon juice for a tangy taste. Eat with a spoon. *Benefit:* Yogurt is a *fermented* product, which means it is a powerhouse of living enzymes. These join with the vegetable enzymes, then melt with the nutrients in the nuts and create a dynamite "pep-up" that makes you feel great. This Enzyme Pep-Up Elixir also staves off hunger pangs as the enzymes take the unsaturated fat from the nutmeats and use the essential fatty acids to promote satiety of your appetite. It's a powerhouse of slimming in a cup of natural foods found almost anywhere for a few pennies!

18. For a tummy-warming and satisfying meal, make a bowl of bouillon soup, then add chopped raw vegetables and natural herbs or spices. With this, have a few rye crisp

crackers or use a few slices of cheese. This gives your enzymes the needed ingredients to control your appetite and also gives you nourishment.

19. When eating vegetables, try to eat them *raw*. While some vegetables must be cooked, nearly all can be eaten raw. The benefit here is that they contain a treasure of enzymes when raw, give you chewing satisfaction and fill you up without fattening you!

20. Enzymes work better when you chew *slowly, thoroughly,* and *leisurely*. Ingested food is better metabolized (less fat buildup) when chewed thoroughly. Simple, yet highly effective—thanks to enzymes.

21. For enzyme-high "lunches in a glass," try these: *Enzyme Slim Drink:* Mix together a cup of soy or skim milk, several sun-dried fruits and a banana. Whiz through a blender. Drink for a treasure of vitamins, minerals, enzymes and protein. Helps metabolize other foods and also serves as a satisfying "lunch in a glass." *Low-Calorie Sweet Surprise:* Blend together one egg yolk, two tablespoons honey, one cup of fresh fruit juice and one teaspoon of any vegetable oil. Include some nuts and seeds. Whir through a blender. Drink as a powerhouse of enzyme-energy that is as low in calories and deliciously sweet as it is healthful.

22. For a satisfying low-calorie but high-enzyme meal, toss together any combination of vegetables with chopped nuts and seeds. Sprinkle some herbs on top. Make a salad dressing of oil and lemon juice. Now eat and enjoy the sheer goodness of this raw food. You'll also be sending a stream of fat-melting enzymes throughout your system. It's the natural way to enjoy good food while keeping healthfully slim.

23. To help take the edge off any persistent appetite, eat a handful of *raw* nuts and seeds. Since many of these are up to 50 percent oil, your body enzymes can use this liquid content to help satisfy your appetite and give you the feeling of satisfaction you need to make you enjoy fullness without adding pounds.

24. To give you needed bulk, chew fibrous vegetables

such as carrots, celery, radishes, lettuce, etc. Enzymes will use these to create the "bulk" you need for regularity as well as a comfortable feeling of fullness.

25. Once a week, go on a raw food fasting program. This is high in enzymes and nutrients and low in fat. It's the healthy way to cleanse your system while you slim down and enjoy a youthful body and mind.

Yes, you can enjoy reducing while you eat with some of the preceding tips, and benefit from the enzymes that help metabolize foods to keep them from turning into excess fat.

**How Lottie G. Used Enzymes to Change from Pleasingly Plump to Slimmingly Sweet.** After her children married and left home, Lottie G. felt she was no longer needed. To compensate for the feeling of loneliness, she started nibbling and eating. It did not take her long to gain weight. She was more than just "pleasingly plump." She was a "plump pig" as someone snidely whispered in giggles behind her back in the dress shop where she had to select "stout" styles. She decided to do something about it. She was advised to control her *urge* to eat with enzymes. Here's how it works:

**Raw Vegetables Soothe and Pamper Appestat.** Enzymes in raw vegetables are sent into the disordered appestat mechanism where the urge to eat begins. Enzymes enter the cells in the midbrain where this appestat is found. Enzymes then use the transported vitamins, minerals and essential fatty acids to soothe, pamper and satisfy this appestat. Once this happens, the urge to gorge on food begins to waver. Enzymes help control this urge. Raw vegetable enzymes also soothe the hypothalamus in the midbrain, which is also involved with the eating urge.

Lottie G. would eat a whole plate of raw vegetables whenever she felt hungry. Seasoned with herbs, a bit of safflower oil and a few drops of lemon juice, and she was able to satisfy her eating urge; at the same time, this powerhouse of enzymes soothed her hypothalamus and appestat. She found herself slimming down because she was able to nip the eating urge to the bud—in her midbrain—by using raw vegetable enzymes.

**Slims Down Curvaceously.** This simple program helped Lottie G. eat less, slim down, and gain such a curvaceously slim figure, she was soon buying dresses in the "Young Miss" section of the department store. Raw vegetable enzymes satisfied her appestat and her frustrations were eased. To make herself feel useful, Lottie G. holds a part-time job. Now she feels life has a youthful purpose—and she looks the picture of youth as living proof—thanks to enzymes!

**How to Make Your Own All-Natural Liquid Diet.** Low-calorie beverages have their benefits in that they do help you control calorie intake. But many are made with synthetic ingredients containing sugar, are void of enzymes, and are little more than chemical drinks! You may lose weight on a liquid diet, but you'll also lose a lot of valuable nutrients. With simple ingredients, you can make your own all-natural enzyme- and nutrient-high drink:

*All-Natural Enzyme Diet Drink:* In a blender (or using a bowl and a wire whisk) combine one cup of fresh fruit juice, a tablespoon of any seed oil, and one tablespoon of a protein powder. You can obtain protein powder at most health food stores and larger supermarkets. In your blender, whiz together a few moments. (If using a bowl, beat with a wire whisk until thoroughly blended.) Then drink. You have about 200 calories or less in one glass. But—you have a powerful supply of vitamins, minerals, proteins, essential fatty acids, and enzymes. All are metabolized at a slow rate to give you nourishing fullness while you slim down. This is the all-natural way to keep slim the enzyme way. Make your own diet drink and improve your health at the same time. It's delicious too.

To help burn up fat, calories, and protein, look to the natural way to metabolism—enzymes. These magic substances promote better body metabolism, help create the "burning" process that turns food into usable nutrients rather than excess fat. You can enjoy eating wholesome, natural foods while you melt pounds with enzymes.

*Essential Points:*

1. To keep slim, give yourself a steady supply of Nature's own reducing agents—enzymes. Follow the simple five-step plan that is outlined.

2. No need to count calories when you follow as many

of the 25 listed "eat and reduce" ways that use enzymes for slimming down. All are natural and effective.

3. For an enzyme-high "lunch in a glass" that is also low in calories, (but high in nutrition) try either the *Enzyme Slim Drink* or the *Low-Calorie Sweet Surprise.*

4. To change from "pleasingly plump" to "slimmingly sweet," Lottie G. used raw vegetables to soothe her appestat and control her runaway appetite. Enzymes turned the tide in her slim favor.

5. Make your own low-calorie but high-enzyme nutritious Liquid Diet. Puts a stopgap on your appetite. *Nourishes and satisfies while it slims you down.*

# 18

# A Treasury of Enzyme Health Secrets for Youthful Living

Simple but effective food enzymes have a treasure of healing benefits. Since enzymes control just about every living function from the power to blink an eyelid to the power of metabolizing tough meat, they offer a "head to toe" healing potential when used in the proper manner.

Here is a collection of treasured enzyme health secrets that help promote more youthful health throughout your body and mind.

## *Ten Youth-Building Benefits of Enzymes in Raw Seeds and Nuts*

Raw seeds and nuts offer a treasure of natural enzymes, as well as a Nature-created type of oil that contributes to the health and youth of your body. When you eat enzyme-rich raw seeds and nuts, these miracle workers use the oils to perform ten basic youth-building benefits:

1. *Enzymes Create Energy.* Enzymes use seed and nut oils to produce a steady flow of energy that is reportedly higher than a comparable amount from protein or carbohydrates.

2. *Enzymes Offer Padding.* Enzymes use seed and nut oils to serve as padding around the vital body organs to help absorb shock and to protect against injury during physical exertion and movement.

3. *Enzymes Offer Heat Conservation.* Enzymes use seed and nut oils to help conserve body heat by preventing loss through the body's eliminative channels. During cold weather, the enzymes use these oils to help promote a steady supply of warmth from within your body.

4. *Enzymes Help Transport Vitamins.* Enzymes use the seed and nut oils to carry Vitamins A, D, E and K throughout your body to help boost your health maintenance. These vitamins are "fat soluble" which means that they must have enzyme-distributed fat (as from seeds and nuts) in order to be distributed and assimilated throughout your body.

5. *Enzymes Offer Satisfaction.* Enzymes use the seed and nut oils to give you a satiety value in meals. Foods that "stick to the ribs" have a "filling" or "satisfying" quality because enzymes use these seed and nut fats to create a slow and more satisfying digestive process. Enzymes use these seed and nut oils to help keep you feeling fuller and more satisfied for a longer period of time. This is a natural appetite control to help protect you against overeating or compulsive eating.

6. *Enzymes Offer Intestinal Lubrication.* Enzymes use the seed and nut oils to serve as an intestinal lubricant to moisten waste and help you establish normal and laxative-free elimination.

7. *Enzymes Help Build Body Cells.* Enzymes use the seed and nut oils to take up phosphorus, a mineral that forms a phospholipid substance that is used to build billions of body cells and tissues from head to toe.

8. *Enzymes Soothe the Nervous System.* Enzymes use the seed and nut oils to form the myelin sheath which coats the nerves, insulating and protecting them from abrasion. Enzymes need the fat from seeds and nuts to make this protective nerve-soothing coating.

9. *Enzymes Distribute Essential Fatty Acids.* Enzymes use the seed and nut oils to nourish the body with the three essential fatty acids, known unofficially as Vitamin F. These are arachidonic acid, linoleic acid and linolenic acid. Enzymes use these three Vitamin F substances to promote internal growth and regeneration of body cells, tissues, organs and

other vital functions. Your body cannot synthesize these essential nutrients and must depend upon enzymes that metabolize them from raw and nonprocessed foods such as seeds and nuts.

10. *Enzymes Offer Good Food Flavor.* Enzymes use the seed and nut oils to improve the flavor and texture of many foods. Whether you eat raw seeds and nuts or use their cold-pressed, nonhydrogenated oils, you will reap a harvest of delicious taste and natural flavor out of any simple meal. Enzymes use these oils to give you needed taste satisfaction and good food enjoyment.

**How to Eat Enzyme-Rich Seeds and Nuts.** As a replacement for bleached sugary confections, have a plate of raw seeds and nuts. For an after-dinner snack, have a plate of assorted raw seeds and nuts. Use raw seeds and nuts in a large raw vegetable salad for a super-plus enzyme value. Sprinkle chopped nuts over a cooked meal just before serving. When travelling, carry along a supply of raw seeds and nuts for a healthful snack that helps keep you feeling youthfully healthy. Between meals, enjoy a handful or two of raw seeds and nuts. Mix them up in a variety or eat them singly, depending upon your individual taste. Be sure to use the cold-pressed oils from raw seeds and nuts as part of a salad dressing for more enzyme-building youth benefits. Nature has endowed the seed and nut with a powerhouse of youthful elements. Feasting upon raw seeds and nuts is as healthful as feasting upon Nature itself!

### *Nature's Antacid Tonic*

A seed has within itself a kernel of self-perpetuation. Nature has placed within a seed almost all the life-giving elements required to sustain itself. In particular, alfalfa seeds contain a treasure of nutrients that are unique. They contain Nature-balanced elements that are soothing to the digestive system and also help ease problems of acidosis. Here is how to use the simple alfalfa seed to create a soothing stomach tonic. It is considered Nature's Antacid Tonic.

*How to Make It:* Place two tablespoons of alfalfa seed (the organic variety is available at all health stores, organic gardening outlets and herbal pharmacies) in a quart jar. Fill this jar half full with fresh water or bottled natural spring water. Let it stand overnight. The following morning, pour off the water into a jar or glass. Mix it with some fresh fruit juice and drink it slowly.

*Benefit:* The enzymes from the seeds have multiplied the vitamin-mineral content and promoted a natural plant oil base. This mixture is now part of the water. You now have an all-natural antacid tonic with plant oils that soothe the ruffled digestive tract, coat the delicate organs with a soothing "envelope" to relieve the burning-churning sensation. Just one glass in the morning and your digestive system will be grateful for the healing relief from acidosis.

*Suggestion:* Do not discard the soaked alfalfa seeds. Instead, rinse under fresh water. Now put them in the jar and cover the mouth with several thicknesses of cheesecloth. Use a rubber band or string to tighten it. Tilt the jar slightly so that whatever moisture has remained can evaporate (but the seeds still need oxygen–hence the use of cheesecloth). Rinse the seeds several times each day. By the end of the third day, the alfalfa seeds will have sprouted bright green leaves. Enzymes have increased their vitamin-mineral-protein content. They are ready to eat.

Eat the whole sprout, including the seed, shell, mineral-rich root, vitamin-rich stem, protein-rich leaves–everything. You will be giving your body a powerhouse of nutrition from Nature's miracle of you–seed sprouts.

*Tip:* Orientals use seed sprouts in scrambled eggs and omelets; they add soy sauce and call it *egg foo yong.* Try it. You'll enjoy it!

### *Raw Foods Promote Internal Self-Cleansing*

The victim of a "nervous stomach," Helen N. would regurgitate much of her food. She became malnourished because of the limited amount of nourishment her body was getting. Furthermore, this nervous housewife felt a "burning and churning" in her abdomen that was complicated by irregularity. She had gone the route of habit-forming stomach tonics and milk-flavored compounds that she called "chalk flavored." It only worsened her condition.

**Easy Enzyme Secret Gives Easy Digestive Healing.** Helen N. listened as a neighbor told how she had paid up to a thousand dollars for a vacation at a health spa, where fresh raw vegetables were served *before* each meal. The neighbor said she felt such digestive healing that she could eat with the gusto of a youngster. If this was the "secret" that cost the neighbor a thousand dollars, Helen N. decided to try it at home.

**Raw Vegetable or Fruits Before Each Meal.** Before she would start any meal, Helen N. had a large raw salad. Enzymes in these raw foods performed two digestive-regenerating benefits: (1) Enzymes create a self-digestion action within the intestinal tract, thus sparing the work of the digestive glands and soothing an overworked system. (2) About 80 percent of the enzymes reach the colon in an unimpaired state by a method known as oxygen fixation. This establishes an anaerobic condition (regeneration in the absence of oxygen) in the intestinal tract. It is this environment that helps the growth of beneficial coli bacteria which then drive away the pathogenic (disease-causing) ones.

Both of these functions are created when a meal is begun with raw vegetables or fruits. Helen N. followed this simple program and within two weeks she felt so healed of the burning sensation in her stomach that she could enjoy healthful food—and she looked healthy too. Now that she could eat without regurgitating, she let natural raw foods and other healthful foods perform the internal rebuilding.

Helen N. used this enzyme secret that had cost her neighbor a thousand dollars—but it was absolutely free to her, except for the pennies-per-serving cost of raw vegetables or fruits. All she had to do was start each of her three daily meals with this raw enzyme plant healer: a secret that gave her digestive health and the feeling of youth!

**Self-Disinfectant.** Nonprocessed blueberry juice contains enzymes that reportedly have a disinfectant benefit upon the intestines. It is helpful in situations of stomach upset. Just sip one glass slowly and feel the sheer goodness of the enzyme action as these miracle workers send nutrients to help correct the internal imbalance and help promote youthful cleansing and a sparkling clean abdominal tract.

### *How to Use the Papaya for Magic Healing*

Legend has it that Spanish explorers discovered the papaya during the 1500s as they journeyed throughout Mexico. These

Spanish soldiers ate a slice of this golden melon and experienced relief for their gastric distress. Throughout the centuries the papaya has been praised for its health-building benefits.

*Skin Problems:* Apply a slice of freshly cut papaya to a blemish and bind with a piece of cotton. Let it remain overnight if possible. This reportedly helps keep skin soft, lighten freckles, and remove aging spots and blemishes.

*Better Digestion:* For stubborn digestive disorders, eat a raw papaya both before and after the meal.

*Benefits:* The papain enzyme reportedly works in acid, alkaline or neutral media. The papaya enzyme is unique because it helps metabolize all proteins, fats, and carbohydrates. No other enzyme has this function.

*Miracle Healer:* Papaya enzymes reportedly[1] helped heal infected open fractures, osteomyeletis, encysted suppurations, and fistulas. *Tip:* Eat papaya regularly to give your body the supply of needed healing enzymes.

**Magic Rejuvenation.** Drs. Hwang and Ivy reported[2] that the papain enzyme has magic-like rejuvenation powers of healing. This fruit reportedly is useful as an anthelmintic (destroys intestinal worms). It also helps relieve chronic diarrhea in youngsters. It is soothing and promotes healing in disorders such as splenomegaly (enlarged spleen) and hepatomegaly (liver enlargement). The doctors report that the papaya can help correct infertility if the cause is traced to endocervicitis (inflammation of the mucous membrane of the cervix. These are just some of the magic rejuvenation benefits from the enzymes of the papaya—a luscious fruit available almost anywhere, with a powerhouse of healing.

**Soothes Ulcers, Wounds, Allergies.** Papaya enzymes reportedly help emulsify waste matter. It is used to soothe ulcers, to help heal infected wounds. The papaya enzymes also help loosen exudate in an inflamed trachea and in disorders of the bronchial tubes. Allergies are often relieved with papaya enzyme treatment. The papaya enzyme is unique in that it helps the body dissolve the scirrhus or encrusted waste tissues that predispose to illness. The papaya enzymes emulsify these wastes and help slough them off the infected body part and out of the system. This helps the body recuperate and regain its health.

---

1. *Presse Medicale,* 64:72, Paris, France.
2. "Enzymes and Healing." *Annals of the New York Academy of Science,* 54: 161-207

**Heals Hemorrhoids.** A 52-year-old woman was reported[3] to have a condition of painful external hemorrhoids. Phillip J. Pollack, M.D. prescribed the usual therapy that included bed rest, sitz baths, and rectal suppositories. But the woman still suffered from this painful and embarrassing problem. Dr. Pollack then prescribed papain (one tablet to be taken every four hours). Results? Within 48 hours, she showed "marked improvement." Her pain and the accompanying swelling began to diminish. Three more days and she was diagnosed as having a complete recovery. The woman did not need to have surgery—thanks to the prescribed use of papain tablets. (This papaya enzyme is available in tablet form at many health stores and herbal pharmacies. Your doctor will prescribe its use for your diagnosed condition.)

**Soothes Slipped Discs.** A hoped-for natural way to heal slipped discs involves the doctor-administered injection of *chymopapain.* (Discs are cartilage cushions that appear between the vertebrae. They act as shock absorbers. If a disc slips, it presses out against the nerves and causes pain.) A doctor reports[4] that he has administered chymopapain to more than 25 persons afflicted with slipped discs with good results. David Teperson, M.D., reports that papaya enzymes "digest" the middle of the protruding spinal disc. This helps to decompress the pressure point involved in causing the pain. Chymopapain enzymes are believed to reduce the inflammation-causing reaction of histamine. The doctor states, " . . . patients wake up without the pain they have endured for years. This could be the antihistamine effect of the enzymes going to work on the long-standing inflammation." It is further reported that close to 5000 persons with slipped discs have been treated with this papaya enzyme treatment which offers hope for a natural way to heal such disorders.

*Suggestions:* To help protect yourself against disorders, eat papayas regularly. Many stores sell frozen papayas, sun-dried papayas, papaya juices and even papaya teas. Feature papaya in your enzyme enrichment program and you can eat your way to better and youthful health.

**More Pep, More Youthful Vigor.** A powerhouse of energy-building enzymes is found in Sprout Milk. To make this, just blend one cup of sprouted wheat (or any grain) with one cup of water.

---

3. Philip J. Pollack, M.D., "Hemorrhoids," *Current Therapeutic Research,* 5:62.
4. David Teperson, M.D., Newstory; *Miami Herald* (6:19-72).

Add a little honey or unsulphured molasses or pure maple syrup for added taste and better nutrient balance. Drink it as a milk substitute. But it does more than serve in that capacity. It gives you a dynamite supply of powerful enzymes that supercharge your sluggish nutrients and help alert your glands. Sprout Milk is an *Enzyme Energy Elixir* that gives you "youth power" together with "go power."

**Carrots for Body Healing.** Fresh raw carrots as well as their freshly prepared juices offer a naturally balanced combination of enzymes, vitamins and minerals that reportedly help nourish the entire body. Enzymes in raw carrot juice act as a natural solvent of wastes. The enzymes use the carrot's supply of Vitamin A to help resist infection. The enzymes use the carrot's mineral supply to help nourish the glands and enter into the secretion of the adrenal hormones. Enzymes use carrot nutrients to soothe the nervous system and help nourish the optic nerve and protect against sight impairment. Enzymes also help neutralize toxic bacterial infections within the organism. The golden carrot, whether eaten raw or in the form of freshly made bubbly sunshine-colored juice, is Nature's storehouse of rich enzymes that help improve your health and fill you with the zest of youth.

Food enzymes hold hope and promise for better health the natural way. Nature has placed them in fresh raw fruits, vegetables, seeds, nuts, whole grains and herbs. Take advantage of Nature's harvest of enzyme health and reap the rewards of abundant youthful health.

*In a Nutshell:*

1. Raw seeds and nuts offer you a minimum of ten enzyme health secrets.

2. Enzymes in the Nature's Antacid Tonic (chemical-free, medicine-free) reportedly help soothe the burning-churning sensations in a tortured digestive tract. It's delicious too.

3. Helen N. soothed her "nervous stomach" and ended the habit of taking "chalk-flavored" concoctions by following an easy enzyme secret of using raw vegetables or fruits before a meal.

4. The papaya fruit offers a treasure of healing benefits reported by many physicians.

5. For youthful energy, try Sprout Milk as an Enzyme Energy Elixir.

6. Carrots offer overall body healing. They provide a delicious way to enjoy youthful health through rejuvenating enzymes in a golden vegetable.

# 19

## *The Simple Way to Be a "Gourmet Enzyme Chef"*

Enzymes are powerful substances in promoting healing, but they are sensitive to certain conditions, notably heat and the presence of chemicals and artificial preservatives. To preserve the power of magic enzymes, here are the important "enzyme laws" of food preparation:

1. *Avoid Heat.* Enzymes are adversely affected by high cooking temperatures, especially those above 120° F. Above that temperature food enzymes are destroyed. Therefore all fruits, vegetables seeds, nuts, grains should be eaten raw. Avoid cooking any food except when it *must* be cooked, as in the case of meats, fish, eggs, and some beans. To preserve enzymes, the foods should be uncooked and unrefined and free from chemical adulterants. Natural, organic foods offer the highest supply of enzymes as well as other essential life-giving nutrients.

2. *Avoid Evaporation.* When you slice open many fruits or vegetables, the invasion of oxygen depletes their valuable enzyme potency, and also lowers vitamin and mineral values in the plant. Therefore if you must cut these plant foods (or use their juices), plan on eating them as soon as possible afterward. This helps give you as much enzyme potency as possible.

3. *Keep Foods Refrigerated.* If you must store foods, then keep them in your refrigerator or freezer. This helps keep their enzymes intact and inactive. Frozen fruits and

vegetables may have a high enzyme content, in contrast to canned fruits and vegetables which have been preheated and processed and therefore devoid of enzymes. If you are unable to obtain fresh fruits and vegetables, select frozen ones for a good enzyme supply.

4. *Cook Foods as Briefly as Possible.* Enzymes are alive and well at temperatures between 32° and 104° F. They become less active at higher temperatures. When cooking reaches 130° F, the enzymes are destroyed. The boiling point of water is 212° F. This means that when you boil fruits, vegetables, seeds, and nuts, all the enzymes are destroyed. Plan to cook foods which are unedible raw as briefly as possible–just long enough to make them edible.

5. *Balance Raw vs. Cooked Foods.* If you must eat cooked foods, balance them with a supply of fresh raw foods. This helps give your body a healthy balance of many nutrients as well as the magic enzymes, without which your nutrients would be inactive and useless.

6. *Eat Whole Grain Foods.* Select cereals and bread foods made with the germ and the inner husk. This calls for the "complete" grain. While baking will destroy enzymes, there is an enzyme benefit in whole grains. Your body has a store of some 700 to 1000 enzymes. These body enzymes are nourished by ingredients in whole grain foods. Furthermore, white flour and refined sugar lead to burdensome pressure on the metabolism and digestive system. These non-foods interfere with your body's enzyme mechanism. This may also lead to excess acidity and injurious response to body enzymes. Eat whole grains so enzymes are able to correct the bacterial flora of the body and promote a youthful and healthy feeling from top to bottom.

7. *Maintain a Simple Protein Balance.* Your body enzymes are proteins themselves. Flatter them with simple protein balance. Emphasize the use of meatless sources of protein from nuts, seeds, plants, whole grains, mild cheeses, yogurt and fermented milks such as buttermilk. This helps complement meat protein and also helps maintain a balance that is soothing to the digestive system.

8. *Use Plant Oils as a Replacement for Animal Fats.* The nonhydrogenated, nonprocessed, cold-pressed plant oils (vegetable, nut and seed oils) are brimming with enzymes. They also enhance the mobility of your body enzymes. Replace hard fats (butter, meat and poultry fats, hydrogenated shortenings, and cream) with these plant oils. This helps balance your body ecology and promote a happy enzyme environment.

9. *Eliminate Enzyme Antagonists.* Non-foods such as refined sugar, harsh salt, pepper, mustard, ketchup, coffee, tea, cocoa, chocolate and any products made from these enzyme antagonists should be eliminated. Replace with healthful herbs and spices, honey, herbal beverages, and coffee substitutes. The enzyme antagonists are not beneficial to the system. Furthermore, they grate and corrode segments of the nervous system and also erode the delicate blood vessels. This is harmful to body enzymes. Your program for healthful living should emphasize natural foods for better enzyme benefit.

10. *Have Raw Vegetables Available.* Always keep a plate of fresh, crisp raw vegetables available. Having this on hand means it is easy to munch your way to better health with natural enzymes.

11. *Try Home-Made Nut Butter.* In a blender, put one cup of any raw nuts (cashews are high in protein, minerals and enzymes together with one tablespoon of sesame seeds and one tablespoon of raw wheat germ. Add 1/3 cup of any coldpressed oil. Mix thoroughly. Blend a few moments. Now refrigerate. Use as an enzyme-rich home-made nut butter on whole grain bread. It is as delicious as it is enzyme-nutritious.

12. *Use Sprouts for Enzyme Enrichment.* To increase the enzyme value of any meal, raw or cooked, add a handful of fresh sprouts just before serving. *Note:* a seed sprout contains more than just many healthful enzymes; it has an increased supply of essential vitamins and minerals. It's the all-natural way to enzyme-enrich any food.

13. *Have a Raw Grain Meal.* You can enjoy softened but raw (enzyme-rich) grains. Here's how: just soak any organic

grains such as wheat, barley, rye, oats, or natural brown rice in a jar of water overnight. Natural bottled spring water is best to use. Next morning, mix the soaked grains with the water, together with fresh raw fruit slices, and chopped or whole nuts and have a raw grain breakfast. It's brimming with nutritional goodness and delicious taste. You'll be sure to remark that enzymes never tasted so good!

14. *Try Sun-Dried Fruits.* Many health stores and large food markets sell sun-dried fruits. These should be clearly marked as "organic" and "no preservatives used." Avoid those that were chemically or artificially dried with sulphuric acid. Natural sun-dried fruits are delicious and also highly enzyme-potent. Eat these sun-dried fruits and give yourself a plentiful supply of needed nourishment. They're also good as a replacement for sugary and useless candies and cakes..

15. *Combine Cooked Foods with Uncooked Enzyme Foods for Total Body Assimilation.* You definitely have to eat cooked foods since many items are inedible unless cooked. This includes the usual meats, fish, eggs, and some vegetables, as well as many beans and grains. To help your body assimilate the nutrients in these cooked foods, balance them with uncooked enzyme foods. The benefit here is that the raw food enzymes will complement the cooked foods to create a better body balance and improved assimilation. Here are some examples:

Combine cooked meats with fresh raw vegetables
Combine cooked poultry with raw bean sprouts
Combine cooked fish with raw nuts
Combine cooked eggs with raw grains
Combine cheeses with fresh fruit
Combine cooked grains with raw grains
Combine cooked vegetables with raw seed sprouts
Combine dairy products with fresh fruit slices
Combine whole grain bread foods with nut butters
Combine soups with seed sprouts
Combine cooked casseroles with raw seeds and nuts

*Benefits:* Since the cooked or processed products (such as dairy and whole grain bread products) are devoid of enzymes, but contain excellent nutritional value insofar as vitamins, minerals, and proteins are concerned, you need to make up for this lack. That is, since cooking destroys enzyme content, you can fortify and enrich the cooked foods the natural way with enzyme-containing foods as listed above. One food *complements* another to create a balance that will give you a powerhouse of nutrients as well as the ever-valuable enzymes. Simple combining does make an enzyme difference. You'll feel it!

You can help increase the magic in magic enzymes with some simple and tastefully rewarding adjustments in preparing foods. Once you discover the flavorful taste of healthful foods the natural way, you'll delight in the delicious way to health rejuvenation, thanks to magic enzymes.

*Highlights:*

1. The 15 natural laws boost flavor and health of foods. They also give you the maximum for a minimum of time as enzymes enter your life and body and boost your youth and health.
2. Combine cooked foods with raw enzyme foods and enjoy better assimilation of many health-building nutrients.
3. Emphasize natural and nonprocessed foods, whether cooked or raw, and you will be taking a giant step toward magic enzyme rejuvenation of your body and your mind.

# 20

## *Your Enzyme Cookbook*

Here's a garden of enzyme recipes that combine good health with good taste. When purchasing any foods, be sure to obtain those that are organic, chemical-free and natural. This gives you the highest potency values of nutrients as well as enzymes. Now—good eating!

### *Blender Seed Drink*

1½ cups of any fruit seeds or vegetable seeds, thoroughly washed
¼ cup dark honey
¼ teaspoon cinnamon
3 cups citrus fruit juice

Use your blender to pulverize seeds until they are powdery. Now add more and more of the liquid until all of it has been used. Add honey, cinnamon and citrus fruit juice. Whir a moment or two. Drink promptly.

### *Apple Delight*

2 cups cored, sliced apples (if organic, do not peel)
½ cup sun-dried fruits (raisins are good)
2 tablespoons honey
¼ teaspoon cinnamon
¼ cup citrus juice

Place the citrus juice and the apples in your blender. Whir until a puree. Add sun-dried fruits, cinnamon and honey. Whir another minute. This makes a delicious applesauce that is brimming with vitamins and minerals as well as enzymes.

### *Enzyme Candy Snacks*

1 cup unsweetened coconut
1 cup sunflower seed meal
1 cup raw wheat germ
¼ cup sesame seeds
½ cup honey

Mix all of the ingredients together. With clean hands, shape them into four or more portions. Then wrap in wax paper. Place in the refrigerator. Slice into snack-like portions as needed. This Enzyme Candy Snack gives you the natural goodness of a confection but without synthetic or chemicalized ingredients. It tastes like halvah.

### *Raw Enzyme Soup*

- 3 potatoes (if organic, do not peel)
- Parsley sprigs
- 3 teaspoons of dry vegetable powder
- 1 sliced onion
- Pinch of sea salt or kelp
- ¼ cup yogurt
- ¼ cup cream
- Chopped chives
- Paprika

Place scrubbed and thinly sliced potatoes in a bowl. Now pour enough boiling water over the slices to cover them. Cover and let stand for five minutes. Now add all the other ingredients. Stir vigorously until completely blended. Before serving, add some more chopped green vegetables for a lovely emerald colour. This is a healthful enzyme soup that is brimming with many other essential nutrients. SUGGESTIONS: To further boost the flavor-taste-nutrition of soups, add some ground peanuts or a few tablespoons of seeds. This creates a powerhouse of enzyme activity that is as healthful as it is good to eat.

### *Fruit Marmalade*

- ½ pound sun-dried apricots
- ½ pound sun-dried prunes
- ½ pound sun-dried peaches or pears

Place all fruits in a deep dish. Cover with enough hot water to come to the brim. Do not cover the dish. Soak the fruit overnight. Next morning, pour off the juice and drink as a treasure of enzymes and nutrients. Now put the soaked fruits through a food grinder. Add enough crushed pineapple and honey to sweeten. Use as a delicious enzyme-high fruit dessert.

### *Seed Topping for Cereals*

- 2 tablespoons ground sesame seeds
- 2 tablespoons ground flaxseed
- 2 teaspoons rice polish
- 2 tablespoons ground sunflower seeds

Just mix together and sprinkle over any cooked cereal. Offers valuable enzymes and essential fatty acids as well as many vitamins, minerals and proteins.

### *Sprout Salad*

2 parts any fresh seed or bean sprouts
1 part diced carrots
1 part cubed avocado or raw sliced beets
Sprinkle of orange or lemon juice

Toss all ingredients together. For a dressing, use equal portions of cold-pressed oil and lemon juice. A high-enzyme salad that can be a meal in itself, with a few slices of whole grain bread and nut butter. Finish with a raw vegetable drink for a powerfully rejuvenating meal.

### *Enzyme Mayonnaise*

1 tablespoon cold-pressed oil
1 teaspoon lemon juice
Grated onion or garlic
2 tablespoons fresh vegetables (diced)

Mix all ingredients together. Stir vigorously to help break up the oil. If possible, whir in a blender. Use as a natural enzyme mayonnaise that is also cholesterol-free.

### *Cheese Nut Ball*

3 cups Danish Bleu cheese, crumbled (about 3/4 pound)
½ cup cold-pressed oil
3 packages (3 ounces each) cream cheese
½ teaspoon paprika
1 tablespoon tomato juice
½ pound walnut meats, coarsely chopped

Blend Danish Bleu cheese with oil. Add cream cheese, seasonings, and tomato juice and blend well. Chill mixture and shape into a ball. Finally, sprinkle thickly with nut meats. Chill before serving. Eat with fresh fruit slices for a high-protein, high-enzyme meal.

*Enzyme Sandwich*

Mix chopped nut meats with seed sprouts and honey. Spread on whole grain bread slices. A delicious sandwich that is full of enzymes and taste.

*Enzyme Butter*

2 cups assorted nut meats
2 tablespoons cold-processed oil
½ teaspoon kelp or sea salt

Grind the nut meats through a food chopper until very fine. Add the oil and sea salt. Mix well until thoroughly blended. Pack in tightly closed containers and refrigerate. Makes a high-enzyme butter that is free of animal products.

*Fruit Soup*

2 cups nut milk
2 cups seasonal berries or fresh fruit slices
2 tablespoons honey
1 sprig mint

Whir all ingredients in your blender until of a smooth consistency. Chill before serving.

*Raw Fruit Cookies*

2 cups pitted organic dates
1 cup sun-dried raisins
Carob powder (a chocolate-tasting substance that is made from beans and sold at most health stores)

Grind dates and raisins together. Add as much carob powder as will hold together. When you have a stiff dough, roll out and slice into pieces. Let it remain in sunshine for a few hours until firm and ready to serve.

# Index

## H



## I

K



L



M



N



O



P



R

Y